Conference Proceedings of the Society for Experimental Mechanics Series

Series Editor

Kristin B. Zimmerman
Society for Experimental Mechanics, Inc.,
Bethel, CT, USA

The Conference Proceedings of the Society for Experimental Mechanics Series presents early findings and case studies from a wide range of fundamental and applied work across the broad range of fields that comprise Experimental Mechanics. Series volumes follow the principle tracks or focus topics featured in each of the Society's two annual conferences: IMAC, A Conference and Exposition on Structural Dynamics, and the Society's Annual Conference & Exposition and will address critical areas of interest to researchers and design engineers working in all areas of Structural Dynamics, Solid Mechanics and Materials Research.

Chad Walber • Matthew Stefanski • Stephen Seidlitz
Editors

Sensors & Instrumentation and Aircraft/Aerospace Testing Techniques, Volume 8

Proceedings of the 41st IMAC, A Conference and Exposition on Structural Dynamics 2023

Springer

Editors
Chad Walber
PCB Piezotronics, Inc
Depew, NY, USA

Stephen Seidlitz
Cummins Power Systems
Minneapolis, MN, USA

Matthew Stefanski
Kettering, OH, USA

ISSN 2191-5644 ISSN 2191-5652 (electronic)
Conference Proceedings of the Society for Experimental Mechanics Series
ISBN 978-3-031-34937-9 ISBN 978-3-031-34938-6 (eBook)
https://doi.org/10.1007/978-3-031-34938-6

This Springer imprint is published by the registered company Springer Nature Switzerland AG
The registered company address is: Gewerbestrasse 11, 6330 Cham, Switzerland

Paper in this product is recyclable.

Preface

Sensors and Instrumentation and Aircraft/Aerospace Testing Techniques represents one of ten volumes of technical papers presented at the 41st IMAC, a Conference and Exposition on Structural Dynamics, organized by the Society for Experimental Mechanics, held February 13–16, 2023. The full proceedings also include volumes on *Nonlinear Structures and Systems*; *Dynamic Substructures*; *Model Validation and Uncertainty Quantification*; *Dynamic Substructures*; *Special Topics in Structural Dynamics and Experimental Techniques*; *Computer Vision and Laser Vibrometry*; *Dynamic Environments Testing*; *Topics in Modal Analysis and Parameter Identification*; and *Data Science in Engineering*.

Each collection presents early findings from experimental and computational investigations on an important area within Sensors and Instrumentation and other Structural Dynamics areas. Topics represent papers on calibration, smart sensors, shock calibration and shock environment synthesis, and applications for aircraft/aerospace structures.

The organizers would like to thank the authors, presenters, session organizers, and session chairs for their participation in this track.

Depew, NY, USA — Chad Walber
Kettering, OH, USA — Matthew Stefanski
Minneapolis, MN, USA — Stephen Seidlitz

Contents

Chapter 1
Modal Analysis of a BattleBot Blade

Aakash Umesh Mange and Chad Walber

Abstract In this chapter, the dynamic characteristics of a BattleBot blade is studied. To optimize the design and improve the structural behavior of a test object, modal analysis is a crucial process. Obtaining modal parameters, namely, the natural frequencies, damping, and mode shapes of the unit under test, facilitates in adjusting the dynamic properties and in improving performance. In this case, a heavyweight BattleBot blade that is designed for combat competitions is tested and analyzed. The results are further used to tweak the design of the BattleBot to make it more resistant to damage or breaking of crucial components. Furthermore, these dimensional changes can also help improve the impact force of the BattleBot's blades when it strikes the opponent. Using a new material for the fabrication can also assist in making the robot lighter and stiffer. Hence, a modal test can help make the BattleBot more efficient for the competitions. This work focuses on the experimental setup and discusses the workflow of modal analysis of the BattleBot blade. A modal hammer is used to excite the blade and a uni-axial accelerometer is used to obtain the vibration characteristics. The short pulse induced with a modal hammer excites a wide range of frequencies. To avoid the mass loading effect that is induced with a roving response measurement, the modal test is carried out with a roving excitation method. The test measurements and results are presented in this chapter.

Keywords Modal analysis · BattleBot blade · Modal parameters · Roving hammer test

1.1 Introduction

Executing modal analysis of a structure is crucial in order to analyze the modal parameters of the test object. The natural frequencies, damping, and mode shapes of the unit under test help in adjusting the mechanical properties of the structure by optimizing the design and improving the structural behavior of the test unit.

Icewave is a heavyweight robot designed for robot combat competitions. Icewave is powered by a 15 HP 2-stroke internal combustion engine from a Husqvarna concrete saw that drives a 54 lb. hardened S7 steel blade, rotating at 200 mph at the tips. This internal combustion engine is used to power the horizontal spinner of the robot. Icewave is also rumored to be the loudest robot because of its engine (Fig. 1.1).

Icewave participates in a competition called BattleBots where the robots fight each other in a combat match. The arena has several obstacles that drivers of the robot can use to damage the opposing robot. Icewave creator and team lead believes some of the losses may have been due to the blade not impacting opponents effectively. Because of the power behind Icewave, some of these losses can be attributed to self-inflicted damage from its own rotating horizontal blade. The modal testing will be used to help increase striking effectiveness.

Examination of the horizontal blade through vibration testing to identify the natural (modal) frequencies could provide useful clues as to how performance might be improved. Such vibration testing is typically performed using a shaker or impact hammer. In this instance, vibration testing was performed with an impact hammer. A quick modal survey shows the natural frequencies, damping, and mode shapes of the robot which helps in identifying the weak links in the structure. Hence, modal analysis is a significant part of optimizing the design and performance of these robots.

A. U. Mange (✉)
Crystal Instruments Corporation, Santa Clara, CA, USA
e-mail: amange@go-ci.com

C. Walber
PCB Piezotronics, Depew, NY, USA
e-mail: cwalber@pcb.com

C. Walber et al. (eds.), *Sensors & Instrumentation and Aircraft/Aerospace Testing Techniques, Volume 8*, Conference Proceedings of the Society for Experimental Mechanics Series, https://doi.org/10.1007/978-3-031-34938-6_1

Fig. 1.1 Icewave competing in BattleBot

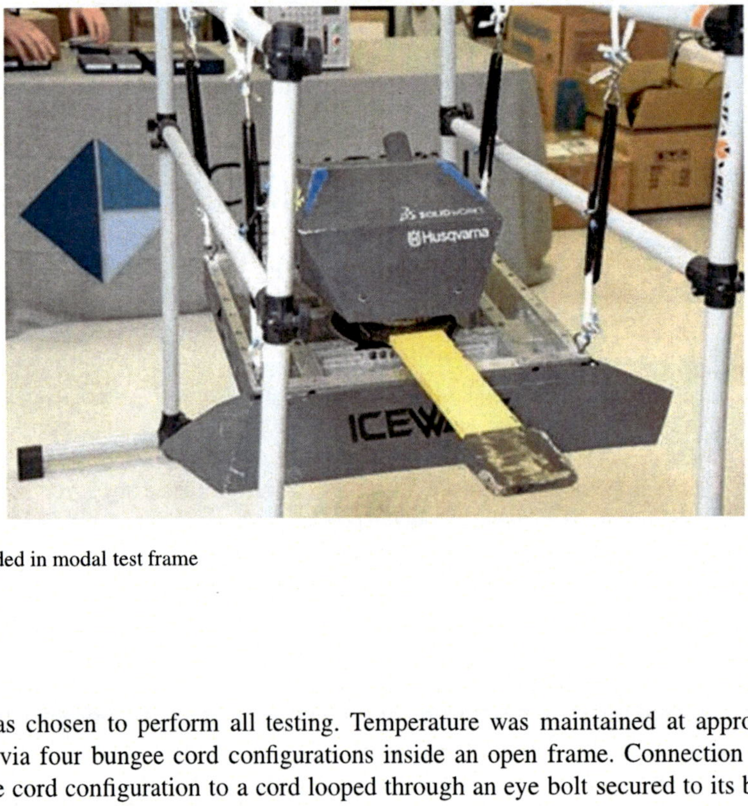

Fig. 1.2 Icewave bot suspended in modal test frame

1.2 Test Layout

An indoor laboratory was chosen to perform all testing. Temperature was maintained at approximately 70 °F (21 °C). Icewave was suspended via four bungee cord configurations inside an open frame. Connection to Icewave was achieved by attaching each bungee cord configuration to a cord looped through an eye bolt secured to its body. The opposite end of each bungee cord configuration was attached to a cord looped over the top rung of the open frame. Thirty PCB Piezotronics ICP accelerometers (Model 333B30) were stud or epoxy-mounted to Icemaker with coaxial cables (PCB Model 002C10) running back to a Crystal Instruments 32-Channel Data Acquisition System (DAQ), Spider 80X with EDM Modal Software. Excitation of the structure was achieved using a PCB modal hammer (Model 086C03).

A hammer impact test is carried out using a modal hammer and a uni-axial sensor to obtain the vibration characteristics of a robot. The short pulse induced with a modal hammer excites a wide range of frequencies. Another advantage of hammer impact test is quick and easy setup. To avoid the mass loading effect that is induced with a roving response measurement, the modal test is carried out with a roving excitation method (Fig. 1.2).

1.3 Equipment Selection

1.3.1 Accelerometers

In modal analysis applications, the general operating requirements are low acceleration signal levels, low frequencies, high channel counts and long cable runs. This requires sensors designed specifically for modal testing with good resolution, low frequency amplitude and phase, small size, flexible mounting, and TEDS (Transducer Electronic Data Sheet).

The above considerations led to select PCB's Model 333B30 accelerometers for this test, as its specifications were a great match. Its low impedance and constant current operation ensures that signal levels are immune to factors that may introduce environmental noise.

Guidelines for selecting accelerometers for modal testing generally state to choose one with 100 mV/g sensitivity with approximately a milli-g resolution in a moderately small package (typically 5 grams or less). This minimizes the mass loading effect of the sensor. Model 333B30 also has a solid track history in modal testing, including similar structures as rudders, helicopter, and turbine blades.

1.3.2 Hammer

PCB's Model 086C03 Modally Tuned® ICP® instrumented impact hammer features a rugged force sensor that is integrated into the hammer's striking surface. "Modal Tuning" is a feature that ensures the structural characteristics of the hammer do not affect measurement results. This is accomplished by preventing hammer resonances in the frequency range of interest from corrupting the test data, resulting in more accurate and consistent measurements.

The force sensor provides a measurement of the amplitude and frequency content of the energy stimulus that is imparted to a test object. Accelerometers are used in conjunction with the hammer to provide a measurement of the object's structural response due to the hammer blow. A variety of tips supplied with each hammer permit the energy content of the force impulse to be tailored to suit the requirements of the item under test.

1.3.3 Modal Analysis System

Crystal Instruments' efficient Spider 80X system along with its powerful EDM Modal software assists in executing the hammer impact modal test.

The Spider-80X analog input channels provide extremely high precision measurements. Each channel has single-ended or differential AC or DC input coupling. It can also provide IEPE (ICPTM) input mode (AC coupling with a 4 mA constant current from a 22 VDC source) for use with industry-standard accelerometers with built-in amplifiers. The ability to read TEDS (Transducer Electronic Data Sheet) identification from the attached transducer completes the channel's compliance with IEEE 1451.4. Each channel provides an unprecedented dynamic range of 160 dBFS, detecting voltages as small as 600 nV and up to 20 V. This is accomplished by applying two 24-bit analog-to-digital converters to each channel and combining their outputs in accordance with our United States Patent number 7,302,354. The Spider-80X also provides time sync Ethernet connectivity and has 4 GB flash memory for data and program storage.

Using multi-channel data acquisition and analysis software, assists to ascertain a variety of mechanical properties leading to an understanding of an object's structural behavioral characteristics. Items analyzed can include resonance detection, mode shapes, transfer characteristics, and structural health via crack and fatigue detection.

1.3.4 Excitation Method

Of special note with modal accelerometers is the consideration of phase. Channel to channel phase matching throughout the measurement system is paramount for modal testing in the global parameter estimation generated from the measured frequency response data base. With global parameter estimation, the consistency of the frequency response function (FRF) database in terms of natural frequencies is also important. For this reason, it is also advised to instrument all desired measurement points simultaneously, thereby providing consistent mass distribution of the sensors on the test structure. Excitation is achieved by moving the impact hammer about the structure under test. This excitation strategy is known as the Roving Hammer Technique.

The older practice of roving a small set of accelerometers about a structure (Roving Accelerometer Technique) runs the risk of shifting certain component resonances as they are loaded and unloaded with the variable mass distribution of the "roved" set of accelerometers. This results in an inconsistent FRF database, which challenges the parameter estimator, as it expects resonances to be consistent, global properties. An additional benefit of instrumenting all measurement points is the reduction of measurement set time. Essentially, the measurement process takes a "snap shot" of data in time, ensuring that other variances (e.g., visco-elastic properties that can change with temperature) are consistent in the measurement database.

1.4 Experimental Setup and Layout

A mesh configuration of 30 measurement points is distributed through the blades to get good spatial resolution for the mode shapes. Using a flexible band and bungee cord, the robot is hung to imitate a free-free boundary condition (as shown in the experimental setup). The modal hammer with the metal tip is roved through the various measurement points. The responses to the impact excitations is captured using a uni-axial accelerometer that is placed accordingly. Measuring the excitation and response in the vertical direction facilitates in obtaining the out-of-plane mode shapes (Fig. 1.3).

For this modal test, the modes up to 3.5 kHz frequency range are of interest and therefore a sampling rate of 8 kHz is set. A block size of 8192 is selected. A fine frequency resolution of 0.9765 Hz is produced with these configuration settings. Measurements of higher accuracy and reduced noise are obtained by linearly averaging three blocks of data at each measurement DOF.

The broadband spectrum from the metal tip on the modal hammer assists in exciting the modes up to a frequency range of 3.5 kHz. The large block size implemented helps in ensuring natural decay of the structure response without introducing the conventional force-exponential window. Another added advantage of this block size is a finer frequency resolution. With this setup, there will be no leakage and a uniform window can be selected (Fig. 1.4).

The coherence plot helps validate the measurement results; it looks good from the above screenshot. The valleys in the coherence plot occur at the anti-resonances, which indicates that the response level is relatively lower at these corresponding frequencies. So, overall, the inputs and outputs are well correlated in the desirable frequency range.

The DOF of the excitation and response for the measured FRF signals are switched automatically for this roving excitation testing. This can be seen from the Modal Data Selection tab.

The FRF measurement shows good dominant peaks in the 0-3500 Hz frequency band. Overlapping the 30 measured FRFs, several modes can be identified. The good alignment of the peaks indicates the measurement results are good and there was no mass loading effect induced (Fig. 1.5).

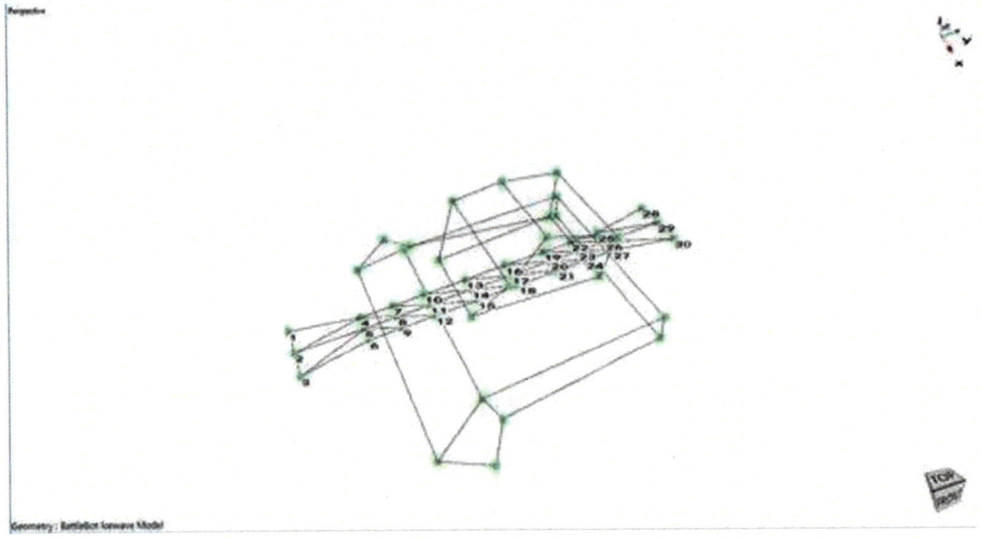

Fig. 1.3 Planned measurement Grid of 333B30 accelerometers

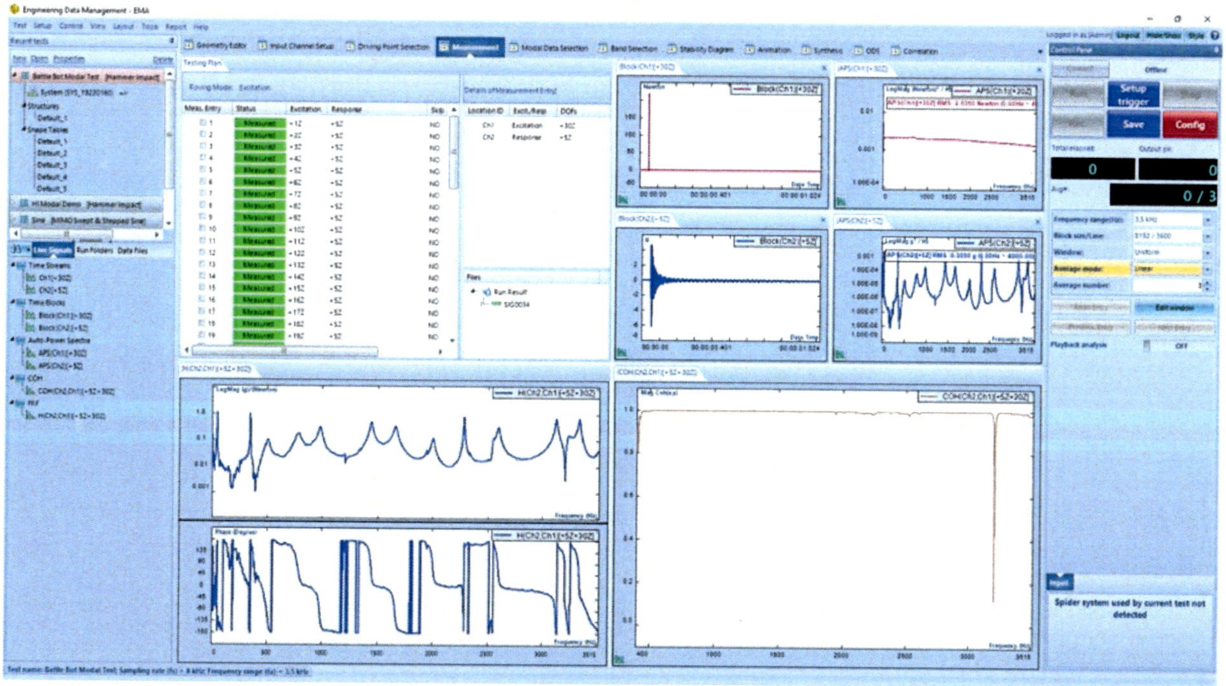

Fig. 1.4 Hammer impact measurement of the robot

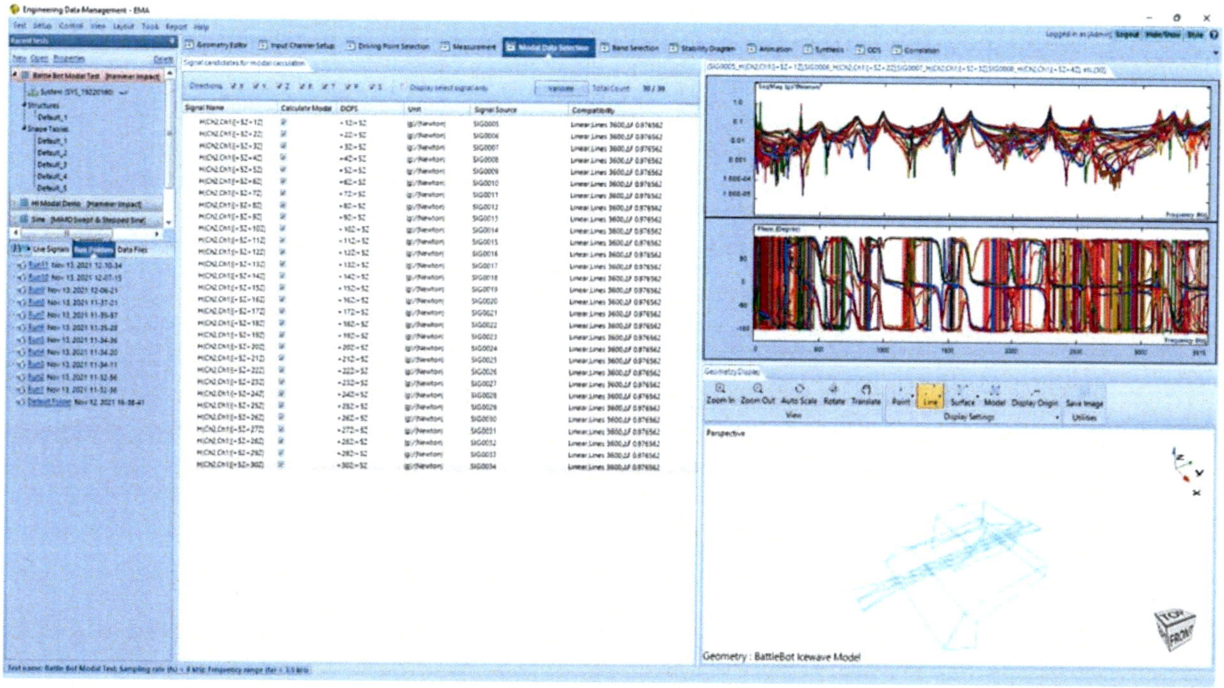

Fig. 1.5 Modal data selection tab showing the magnitude and phase part of all overlapped FRFs

The Complex Mode Indicator Function (CMIF) is used to locate the modes in the desired frequency range. In addition, the summed FRF is also observed to identify the modes. The frequency domain–based **Poly-X** method is used to curve-fit the FRFs to procure the following stability diagram. Six flexible modes are selected within the desired frequency range (Fig. 1.6).

The stable poles (stable frequency and stable damping) are selected to obtain the natural frequencies and the damping ratios of the interested modes. The residue calculation facilitates in obtaining the mode shapes associated for each of the modes (Figs. 1.7, 1.8, 1.9 and 1.10).

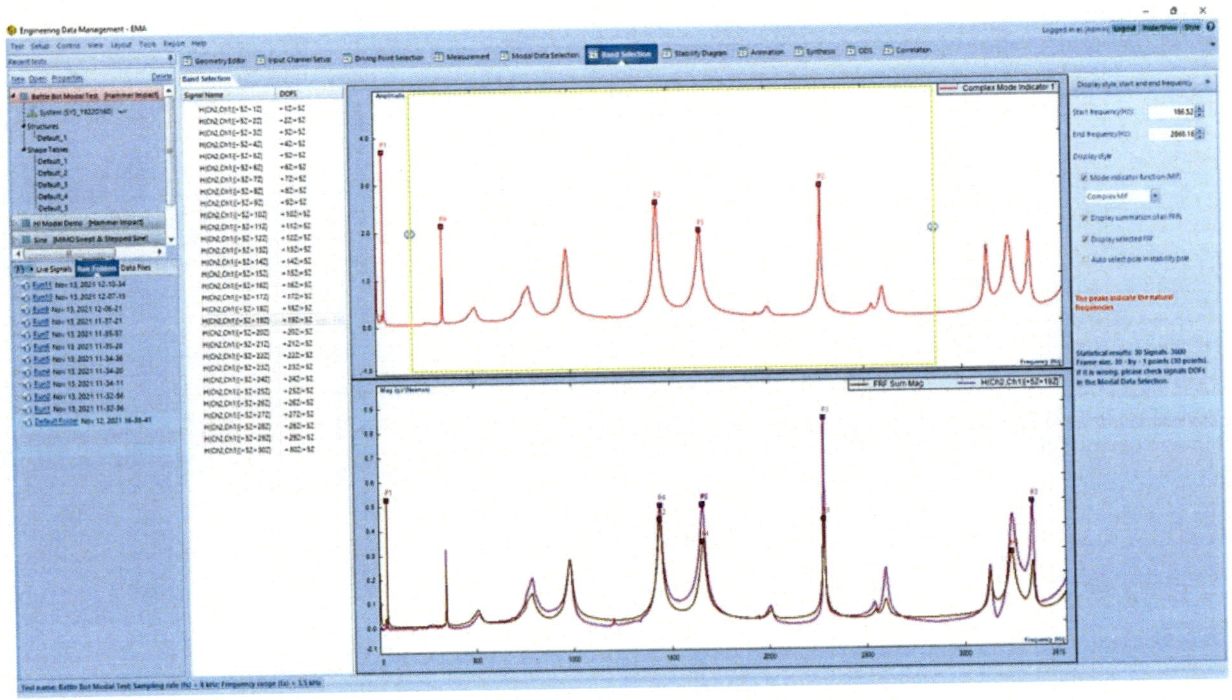

Fig. 1.6 Mode indicator functions to locate and identify the modes in the desired frequency range

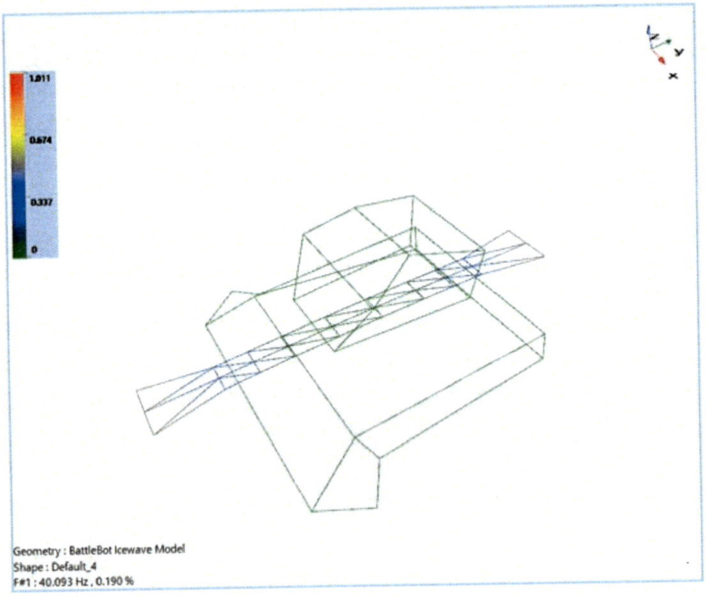

Fig. 1.7 1st order bending mode of the robot at 40 Hz

A quick sweep through the measured FRF dataset can also be carried out to visualize the deformation of the robot at each of the frequencies. With this spectrum data, the modes are uncoupled and hence the operational deflection shapes and the transition through these different frequencies can be analyzed and studied using the ODS function (Fig. 1.11).

1.5 Conclusion

The models suggest combined bending and torsion modes of the striking arm reduces the potential of the destructive force imparted on an opponent.

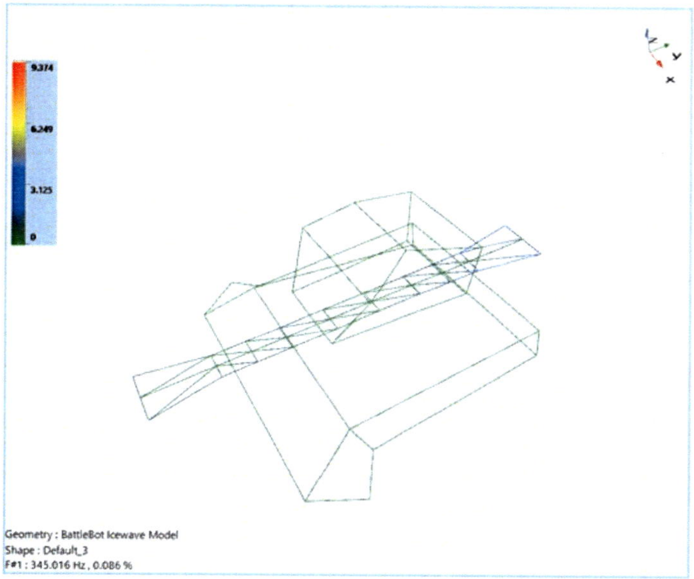

Fig. 1.8 1st order torsion mode of the robot at 345 Hz

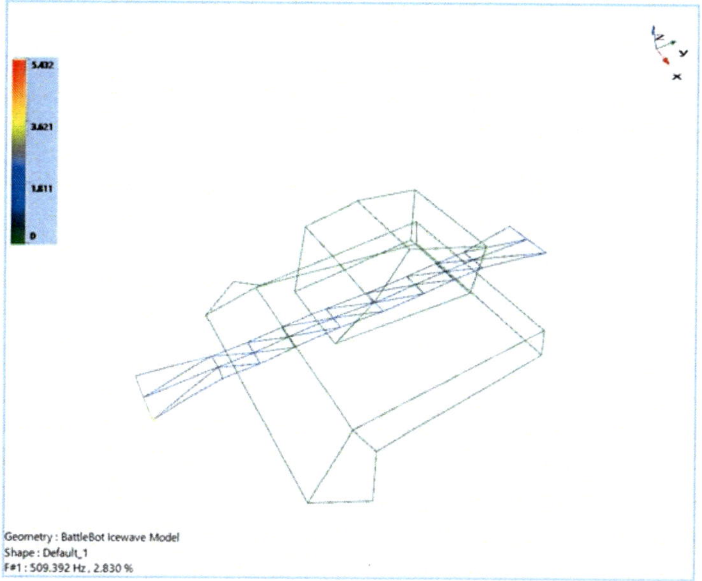

Fig. 1.9 Higher order bending mode of the robot at 509 Hz

Moving forward, the Icewave team will have multiple options to potentially improve striking performance in combat. One general approach would be to look at ways to "stiffen" the striking arm. This could be achieved by increasing the thickness of the arm and accepting any weight penalties or by evaluating new materials for the arm that are less susceptible to torsion and that will keep a better edge.

A third option would involve modifying the basic design of the arm on the bot. Torsion box physics might be turned to, as a lighter arm constructed with ribs will improve stiffness, yet it begs the question if it will deliver the force needed to incapacitate opponents.

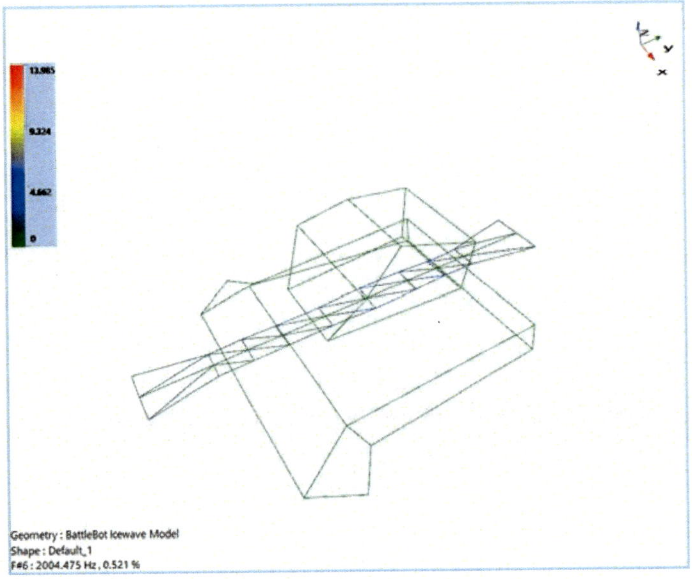

Fig. 1.10 A combination of bending & torsion mode of the robot at 2004 Hz

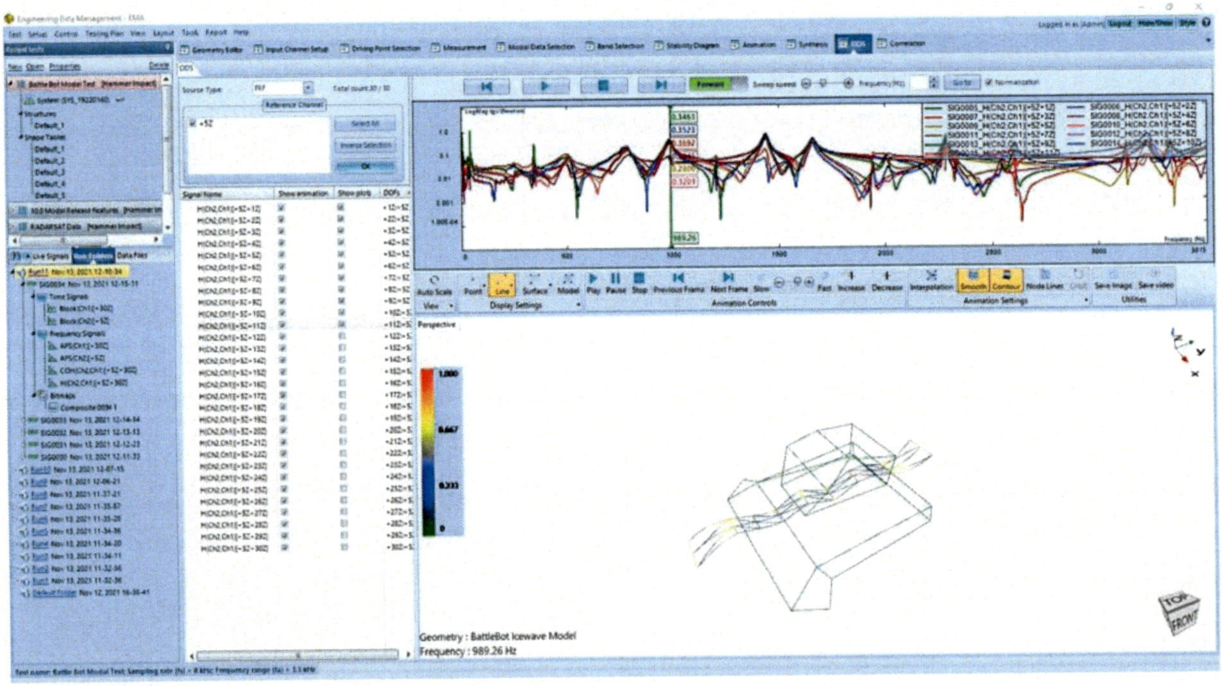

Fig. 1.11 Frequency domain ODS of the robot

Chapter 2
Test Data Measurement Uncertainty Analysis

David T. Buck

Abstract The Structural Validation Branch (AFRL/RQVV) is a research and development test organization that supports internal (Air Force) and external customers. The calculation of test data uncertainty is a necessary capability for a test organization. Test data uncertainty analysis provides error limit values based on the combined effects of random and systematic error sources. The measurement uncertainty calculations provide a means to characterize the measurement methods and understand what the test data represent. This allows engineers to make informed decisions based on the quality of the test data. Error source examination can also be used to determine if corrective actions are necessary and/or possible to eliminate or reduce measurement errors. AFRL/RQVV has completed a project to develop the procedures and tools to calculate and document test data measurement uncertainty values.

Keywords Test · Data · Measurement · Uncertainty · Analysis

2.1 Introduction

The Air Force Research Laboratory Structural Validation Branch (AFRL/RQVV) performs static, dynamic, acoustic, thermal, and vibration testing of aerospace structures. All RQVV tests use instrumentation and equipment that meet the Air Force Metrology and Calibration program. AFRL/RQVV has not provided test data measurement uncertainty values with test data. Instrument accuracy (calibration) was known. However, the total measurement uncertainty value, including uncertainties related to the data collection system, cables and connectors, and the random data scatter based on a statistical analysis of the test data, was not known.

AFRL/RQVV initiated and completed a program to develop the procedures and tools to calculate and document test data measurement uncertainty values. The project had the following primary goals:

1. Review established uncertainty calculation methods and develop "user friendly" procedures for RQVV application.
2. Compile a list of critical instrumentation, and complete an uncertainty analysis for each item.
3. Create an uncertainty database.
4. Develop guidelines for documenting and reporting measurement uncertainty values.

The critical instrumentation list (item 2 above) was a sampling of the most used instrumentation in RQVV test programs. The list included static load cells, axial strain gauges, optical pyrometers, thermocouples, uniaxial accelerometers, and string potentiometers.

The author's presentation at IMAC XL in Orlando, FL (February 2022), provided a background of the measurement uncertainty definitions, calculation methods, project goals, and accomplishments to date. This chapter is intended to provide a short summary of the process and present the final project results, including details for the uncertainty calculations for a uniaxial accelerometer.

D. T. Buck (✉)
Aerospace Engineer, Air Force Research Laboratory, Aerospace Systems Directorate, Aerospace Vehicles Division, Structural Validation Branch (AFRL/RQVV), Wright-Patterson, OH, USA
e-mail: david.buck.5@us.af.mil

C. Walber et al. (eds.), *Sensors & Instrumentation and Aircraft/Aerospace Testing Techniques, Volume 8*, Conference Proceedings of the Society for Experimental Mechanics Series, https://doi.org/10.1007/978-3-031-34938-6_2

2.2 Background

Test data uncertainty analysis provides measurement error limit estimates based on the combined effects of random and systematic error sources. The following sections will briefly describe the two error sources and define the method for combining their effects into a final measurement uncertainty value.

Distribution Statement A: Approved for Public Release; Distribution is Unlimited. AFRL-2022-5078, 20 Oct 2022.

2.3 Random Uncertainty

Random errors, also known as precision errors, are calculated with statistical methods. They add scatter to the test data and increase the scatter in the final calculated uncertainty value. Random errors usually follow the data scatter of a Gaussian (normal) distribution. The presence of a random error source usually means there are data to calculate a standard deviation. For data sets greater than 30 points, 68% of the measurements will fall within ± one standard deviation and 95% of the measurements will fall within ± two standard deviations (Fig. 2.1).

The data scatter is defined by the sample standard deviation (Eq. 2.1).

$$\text{Sample Standard Deviation } S_x = \left[\frac{\sum_{i=1}^{N} (x_i - \overline{x})^2}{N-1} \right]^{1/2} \tag{2.1}$$

The final uncertainty calculation requires a term that defines the scatter of the averages. This term is known as random standard uncertainty (Eq. 2.2).

$$\text{Random Standard Uncertainty } S_{\overline{x},R} = \frac{S_{x,i}}{\sqrt{N_i}} \tag{2.2}$$

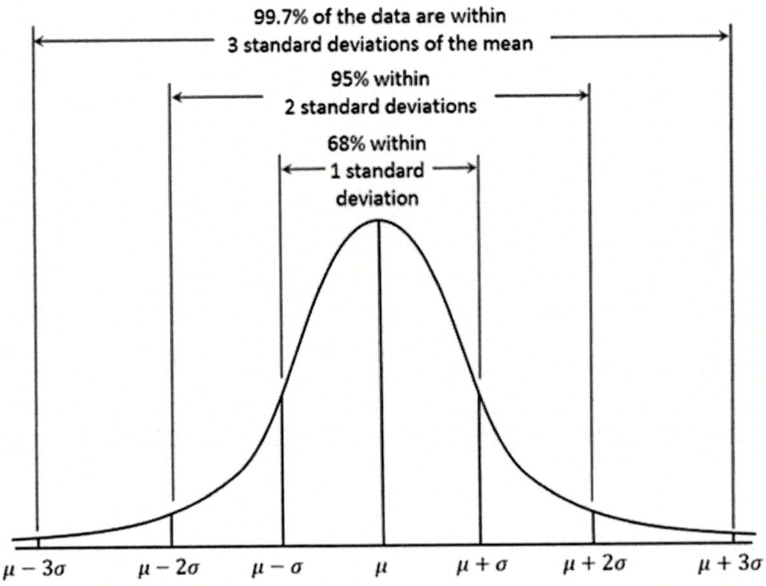

Fig. 2.1 Gaussian (normal) distribution

2.4 Systematic Uncertainty

Systematic errors, also known as bias or offset errors, are usually not estimated using statistical methods and do not add scatter to the test data. Systematic errors are constant for the duration of the test. They are based on given information, such as instrument calibration, manufacturer's uncertainty data, installation errors, and losses through cabling and connections. Manufacturer's uncertainty data are usually given at a 95% confidence level (± two standard deviations).

2.5 Final Uncertainty Calculation

The final uncertainty calculation is performed using Eq. 2.3. This is known as the U_{95} Uncertainty Model.

$$U_{95} \text{ Uncertainty Model } U_{95} = \pm t_{95}\left[(b_R)^2 + \left(S_{\overline{X},R}\right)^2\right]^{1/2} \tag{2.3}$$

Where:

t_{95} = Student's t
b_R = Combined Systematic Standard Uncertainty
$S_{\overline{X},R}$ = Combined Random Standard Uncertainty

The calculated U_{95} value represents the measurement uncertainty at a 95% confidence level. All uncertainty calculations are performed using terms at a 68% confidence level (± one standard deviation). The final measurement uncertainty values are usually reported at a 95% confidence level. A data scatter multiplication factor known as Student's t is used calculate the 95% confidence level from the 68% confidence level. For example, when a sample size is ≥30, the value of Student's t is 2.000. Therefore, 68% confidence level × 2.000 = 95% confidence level.

2.6 Consultant

AFRL/RQVV used the consulting services of Mr. Ronald Dieck (Ron Dieck Associates, Inc.) to provide guidance and assistance in the development and implementation of the uncertainty program. Mr. Dieck is a nationally recognized expert in measurement uncertainty. He has over 35 years of experience in the aerospace industry. He is also the current vice-chairman of the ASME Test Uncertainty Committee (after 25 years as chairman).

Mr. Dieck [1] is the author of the book *Measurement Uncertainty, Methods and Applications*. This book (including the 4th Edition) was used extensively as a reference for the RQVV uncertainty effort.

Mr. Dieck developed the uncertainty calculation procedures, assisted with the development of the database, and performed the initial uncertainty calculations. The primary tool for uncertainty calculations is an Excel spreadsheet that was developed by Mr. Dieck. Random and systematic uncertainty error source values are input at the 68% confidence level along with the associated degrees of freedom. The spreadsheet combines the terms (errors and degrees of freedom) and provides the final uncertainty value at the 95% confidence level.

2.7 ASME and ISO Compliance

The measurement uncertainty calculation methods developed for AFRL/RQVV by Mr. Dieck comply with the following standards:

- American Society of Mechanical Engineers, US National Standard on Uncertainty, ASME PTC19.1-2018, Test Uncertainty, ASME, New York, NY, 2018
- International Organization for standardization (ISO), guide to the Expression of Uncertainty in Measurement (1993)

These two standards define the nationally and internationally accepted uncertainty calculation process. They are very similar; the primary difference being how the error sources and uncertainties are categorized, or grouped, for analysis.

The ASME standard places error sources and uncertainties into one of two error categories: random or systematic. Random errors cause scatter in test results. Systematic errors displace the entire set of test data a fixed amount from the true value.

The ISO guide places error sources and uncertainties into one of two categories: Type A and Type B. Type A uncertainties are calculated using statistical analysis. This usually means that there is data with which to calculate a standard deviation. Type B uncertainties are not calculated using statistical analysis. They are based on given information such as instrument calibration, manufacturer's uncertainty data, installation errors, and losses through cabling and connections.

2.8 Project Accomplishments

A list of critical instrumentation was compiled at the beginning of the project. This list is shown in Table 2.1. An uncertainty analysis was completed for each item. The final uncertainty at a 95% confidence level is shown in the right-hand column. In each case, the systematic error sources were identified, examined, and quantified. Random errors were calculated using test data.

2.9 Uncertainty Database

A Measurement Uncertainty database (Excel format) was created to provide a central location for all calculated uncertainty values and supporting documents. The database contains multiple sheets, one for each type of instrumentation (load cells, accelerometers, strain gauges, etc.). The database information is presented in tabular form. The table contents vary by instrument type but, generally, contain the instrument manufacturer, model or part number, range, operating conditions, installation considerations, and the calculated uncertainty value at a 95% confidence level. Figure 2.2 shows the table that appears on the static load cell sheet. The button labeled "Support Docs" allows access to the available documentation (uncertainty calculation spreadsheets and summary reports) for the individual pieces of instrumentation.

2.10 Detailed Example

The uniaxial accelerometer referenced in Table 2.1 (additional details below) is a common piece of instrumentation in thermoacoustic and vibration tests conducted by RQVV. The final calculated uncertainty at a 95% confidence level was ±15.8% of reading. Figure 2.3 shows the calculation spreadsheet. The yellow arrow at the bottom of the figure points to the cell with the final uncertainty value. Discussion of the spreadsheet entries follow.

The following two systematic uncertainties were identified and used in the final uncertainty calculations.

1. Sensitivity error from the calibration sheet

Table 2.1 Initial uncertainty calculations

Type	Manufacturer	Model or part number	Uncertainty (95% confidence)
Static load cell	Lebow	3143-101-500	± 1.2 lbf
Static load cell	Interface	1010CRA-2.5K-B	± 5.8 lbf
Static load cell	Interface	1010CRA-5K-B	± 4.8 lbf
Optical pyrometer	LumaSense	IMPAC IN 5/9 plus	± 6.2% reading
Optical pyrometer	Williamson	PRO SP-GL-30-F	± 7.2% reading
Uniaxial accelerometer	PCB	252C22	± 15.8% reading
String potentiometer	Vishay	CDS-20	± 0.02 inches
Thermocouple		Type K	± 6.6 °F (68–559 °F) ± [(0.0087 T) + 1.6] °F (560–2012 °F)
Axial strain gauge	Micro measurements	CEA-13-250UN-350	± 11.468 L − 3.5188 L + 1.9422 micro-strain L = Log_{10} of the measured strain

	A	B	C	D	E
1	Support Docs				
2	Manufacturer	Model Number	Full Scale Range, Tension and Compression (lbf)	Test Data used for Calculations	Uncertainty (lbf) 95% Confidence Level
3	LeBow	3143-101-500	0 - 500	P057 DQIV	± 1.2
4	Interface	1010CRA-2.5K-B	0 - 2,500	P158 WiSDM	± 5.8
5	Interface	1010CRA-5K-B	0 - 5,000	P158 WiSDM	± 4.7
6	Interface	1210BF-10K-B	0 - 10,000	P073 MAI Bulkhead	± 28.9
7					
8					

Fig. 2.2 Database sample table

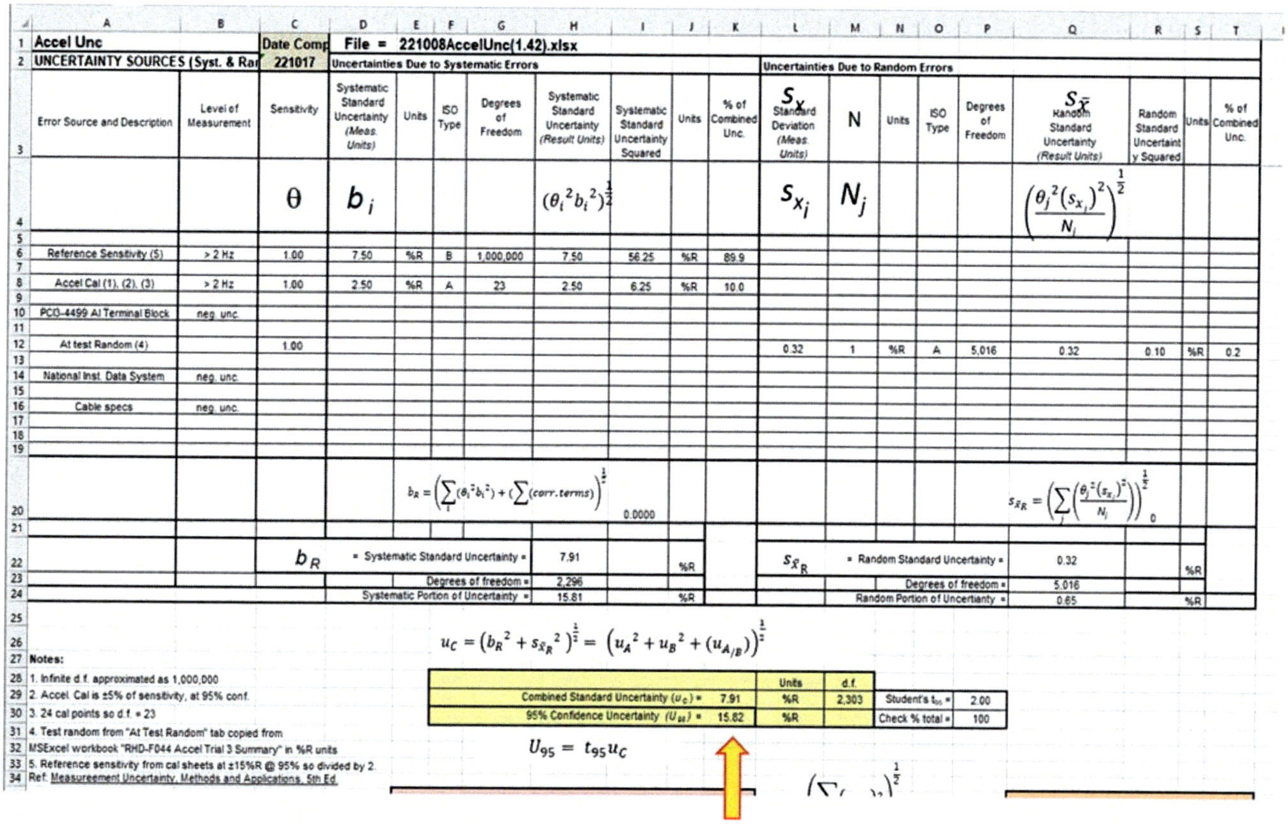

Accel Unc — Date Comp 221017 — File = 221008AccelUnc(1.42).xlsx

UNCERTAINTY SOURCES (Syst. & Ran

	Uncertainties Due to Systematic Errors									Uncertainties Due to Random Errors								
Error Source and Description	Level of Measurement	Sensitivity	Systematic Standard Uncertainty (Meas. Units)	Units	ISO Type	Degrees of Freedom	Systematic Standard Uncertainty (Result Units)	Systematic Standard Uncertainty Squared	Units	% of Combined Unc.	S_x Standard Deviation (Meas. Units)	N	Units	ISO Type	Degrees of Freedom	$S_{\bar{x}}$ Random Standard Uncertainty (Result Units)	Random Standard Uncertainty Squared	Units	% of Combined Unc.
		θ	b_i				$(\theta_i^2 b_i^2)^{\frac{1}{2}}$				s_{x_j}	N_j				$\left(\frac{\theta_j^2 (s_{x_j})^2}{N_j}\right)^{\frac{1}{2}}$			
Reference Sensitivity (5)	> 2 Hz	1.00	7.50	%R	B	1,000,000	7.50	56.25	%R	89.9									
Accel Cal (1), (2), (3)	> 2 Hz	1.00	2.50	%R	A	23	2.50	6.25	%R	10.0									
PCO-4499 AI Terminal Block	neg. unc.																		
At test Random (4)		1.00									0.32	1	%R	A	5,016	0.32	0.10	%R	0.2
National Inst. Data System	neg. unc.																		
Cable specs	neg. unc.																		

$$b_R = \left(\sum_i (\theta_i^2 b_i^2) + \left(\sum (corr.terms)\right)\right)^{\frac{1}{2}} \quad 0.0000$$

$$s_{\bar{x}_R} = \left(\sum_j \left(\frac{\theta_j^2 (s_{x_j})^2}{N_j}\right)\right)^{\frac{1}{2}} \quad 0$$

b_R = Systematic Standard Uncertainty =	7.91	%R	$s_{\bar{x}_R}$ = Random Standard Uncertainty =	0.32	%R
Degrees of freedom =	2,296		Degrees of freedom =	5,016	
Systematic Portion of Uncertainty =	15.81	%R	Random Portion of Uncertainty =	0.65	%R

$$u_C = (b_R^2 + s_{\bar{x}_R}^2)^{\frac{1}{2}} = \left(u_A^2 + u_B^2 + (u_{A/B})\right)^{\frac{1}{2}}$$

		Units	d.f.		
Combined Standard Uncertainty (u_C) =	7.91	%R	2,303	Student's t_{95} =	2.00
95% Confidence Uncertainty (U_{95}) =	15.82	%R		Check % total =	100

$$U_{95} = t_{95} u_C$$

Notes:
1. Infinite d.f. approximated as 1,000,000
2. Accel Cal is ±5% of sensitivity, at 95% conf.
3. 24 cal points so d.f. = 23
4. Test random from "At Test Random" tab copied from MSExcel workbook "RHD-F044 Accel Trial 3 Summary" in %R units
5. Reference sensitivity from cal sheets at ±15%R @ 95% so divided by 2.

Ref: Measurement Uncertainty, Methods and Applications, 5th Ed.

Fig. 2.3 Uncertainty calculation spreadsheet

Uniaxial Accelerometer	
Manufacturer: PCB	Full Scale Range: ± 500 g
Model Number: 352C22	Sensitivity: ~10 mV/g at 100 Hz
Serial Number: 38980	Frequency Range: 0 to 10 kHz

2. Manufacturer's measurement error

Row 6 is the sensitivity error from the calibration sheet. The sensitivity error at 100 Hz is ±15%. Figure 2.4 shows the calibration sheet for one of the accelerometers that was used for the test. The error was assumed to be at a 95% confidence

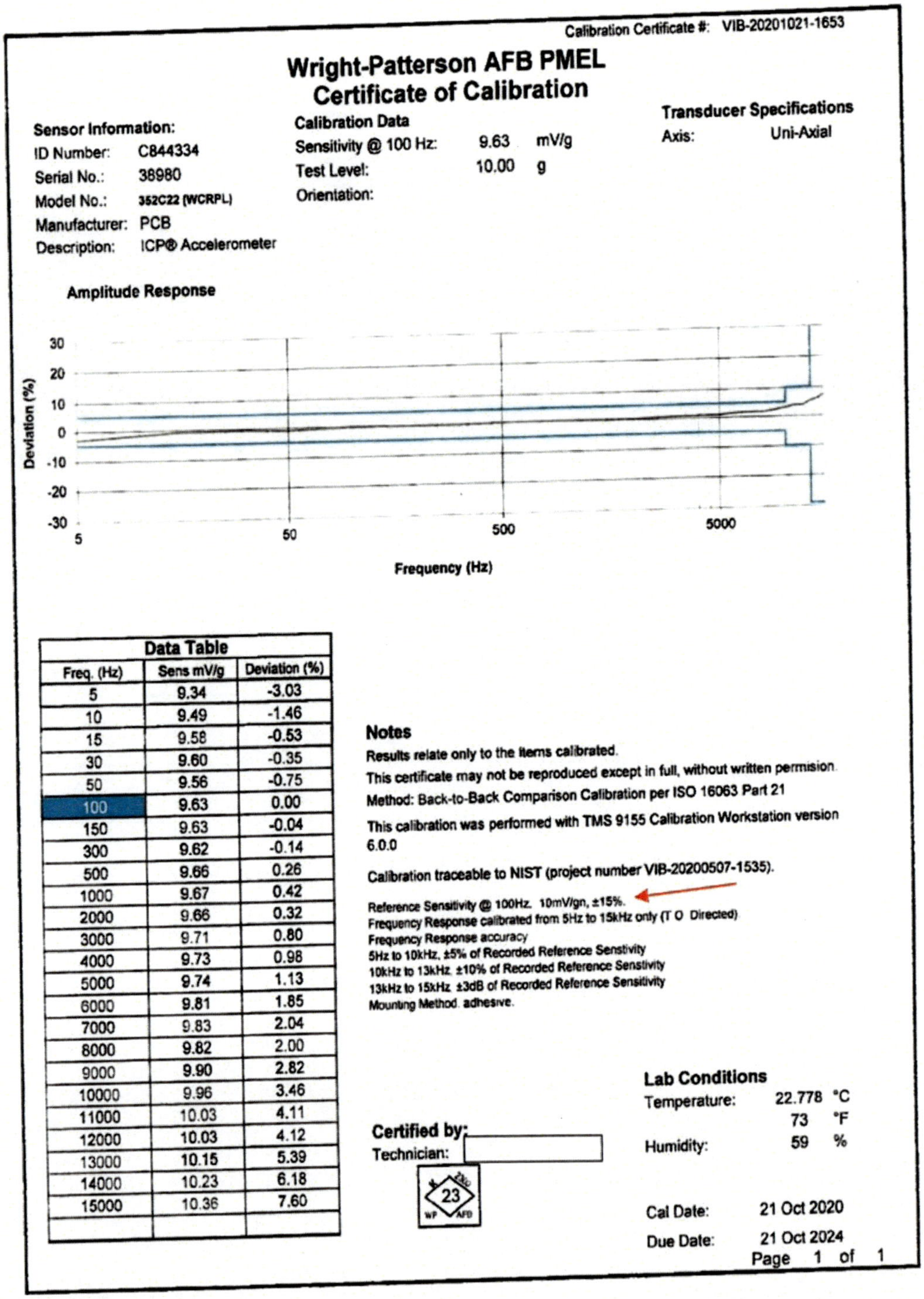

Calibration Certificate #: VIB-20201021-1653

Wright-Patterson AFB PMEL
Certificate of Calibration

Sensor Information:
ID Number: C844334
Serial No.: 38980
Model No.: 352C22 (WCRPL)
Manufacturer: PCB
Description: ICP® Accelerometer

Calibration Data
Sensitivity @ 100 Hz: 9.63 mV/g
Test Level: 10.00 g
Orientation:

Transducer Specifications
Axis: Uni-Axial

Data Table		
Freq. (Hz)	Sens mV/g	Deviation (%)
5	9.34	-3.03
10	9.49	-1.46
15	9.58	-0.53
30	9.60	-0.35
50	9.56	-0.75
100	9.63	0.00
150	9.63	-0.04
300	9.62	-0.14
500	9.66	0.26
1000	9.67	0.42
2000	9.66	0.32
3000	9.71	0.80
4000	9.73	0.98
5000	9.74	1.13
6000	9.81	1.85
7000	9.83	2.04
8000	9.82	2.00
9000	9.90	2.82
10000	9.96	3.46
11000	10.03	4.11
12000	10.03	4.12
13000	10.15	5.39
14000	10.23	6.18
15000	10.36	7.60

Notes
Results relate only to the items calibrated.
This certificate may not be reproduced except in full, without written permision.
Method: Back-to-Back Comparison Calibration per ISO 16063 Part 21
This calibration was performed with TMS 9155 Calibration Workstation version 6.0.0
Calibration traceable to NIST (project number VIB-20200507-1535).
Reference Sensitivity @ 100Hz. 10mV/gn, ±15%.
Frequency Response calibrated from 5Hz to 15kHz only (T O Directed)
Frequency Response accuracy
5Hz to 10kHz, ±5% of Recorded Reference Senstivity
10kHz to 13kHz ±10% of Recorded Reference Senstivity
13kHz to 15kHz ±3dB of Recorded Reference Sensitivity
Mounting Method adhesive.

Certified by:
Technician:

Lab Conditions
Temperature: 22.778 °C
73 °F
Humidity: 59 %

Cal Date: 21 Oct 2020
Due Date: 21 Oct 2024
Page 1 of 1

Fig. 2.4 Accelerometer calibration sheet

level. All spreadsheet entries must be at a 68% confidence level; therefore, the ±15% error was divided by 2 to get the 68% confidence level (± 7.50). The degrees of freedom were assumed to be infinite.

Row 8 is the measurement error from the manufacturer's installation and operating manual. Figure 2.5 shows the page from the manual. The error is stated to be ±5% in the 1 to 10,000 Hz range. Again, this number was divided by 2 to get the 68% confidence level (± 2.5%). The degrees of freedom were assumed to be infinite.

The errors from the National Instruments data collection system and cables (rows 14 and 16) were considered negligible for this analysis. Further investigation of the errors associated with the data collection systems used by RQVV is planned.

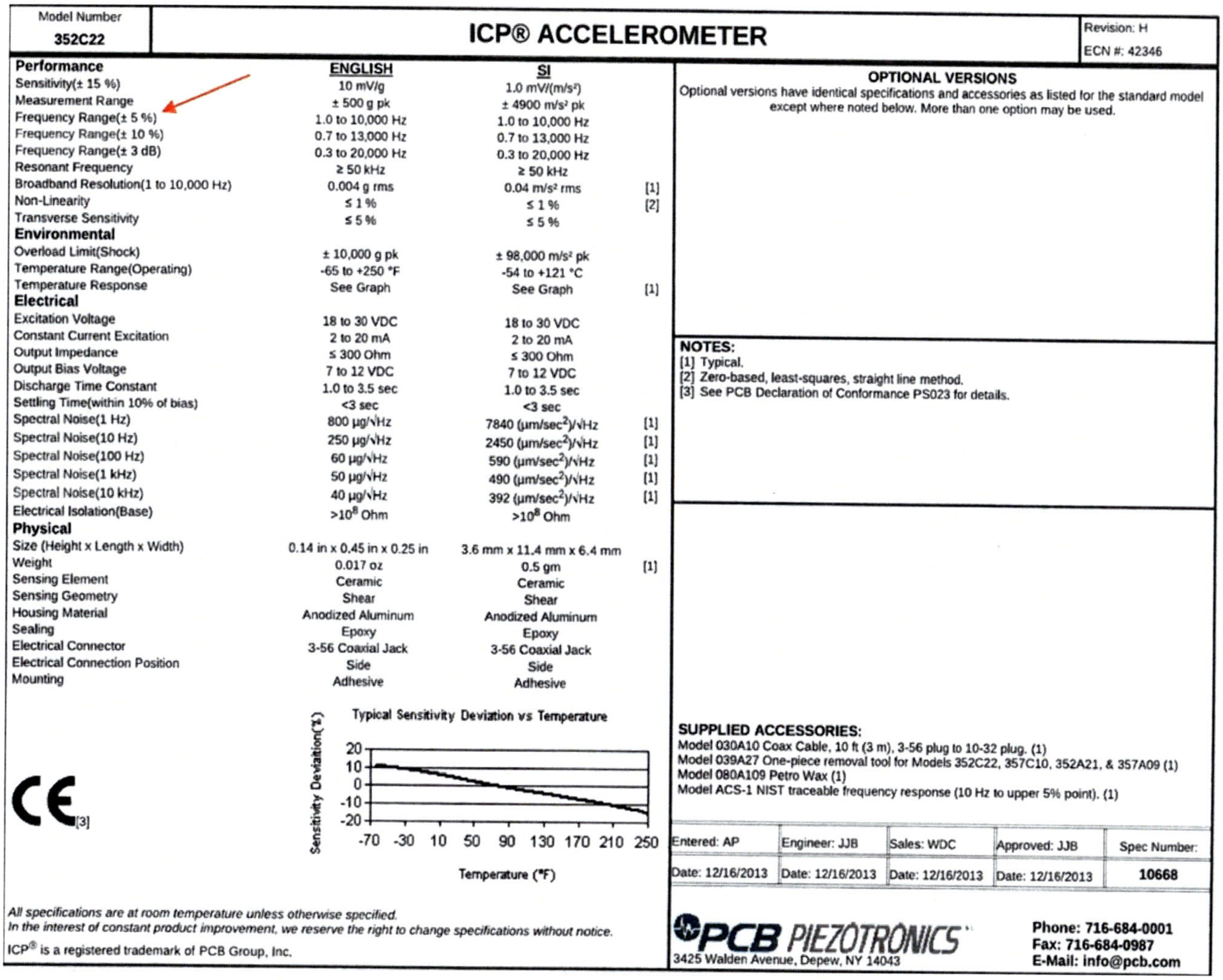

Model Number 352C22	ICP® ACCELEROMETER			Revision: H ECN #: 42346
Performance	**ENGLISH**	**SI**		
Sensitivity(± 15 %)	10 mV/g	1.0 mV/(m/s²)		
Measurement Range	± 500 g pk	± 4900 m/s² pk		
Frequency Range(± 5 %)	1.0 to 10,000 Hz	1.0 to 10,000 Hz		
Frequency Range(± 10 %)	0.7 to 13,000 Hz	0.7 to 13,000 Hz		
Frequency Range(± 3 dB)	0.3 to 20,000 Hz	0.3 to 20,000 Hz		
Resonant Frequency	≥ 50 kHz	≥ 50 kHz		
Broadband Resolution(1 to 10,000 Hz)	0.004 g rms	0.04 m/s² rms	[1]	
Non-Linearity	≤ 1 %	≤ 1 %	[2]	
Transverse Sensitivity	≤ 5 %	≤ 5 %		
Environmental				
Overload Limit(Shock)	± 10,000 g pk	± 98,000 m/s² pk		
Temperature Range(Operating)	-65 to +250 °F	-54 to +121 °C		
Temperature Response	See Graph	See Graph	[1]	
Electrical				
Excitation Voltage	18 to 30 VDC	18 to 30 VDC		
Constant Current Excitation	2 to 20 mA	2 to 20 mA		
Output Impedance	≤ 300 Ohm	≤ 300 Ohm		
Output Bias Voltage	7 to 12 VDC	7 to 12 VDC		
Discharge Time Constant	1.0 to 3.5 sec	1.0 to 3.5 sec		
Settling Time(within 10% of bias)	<3 sec	<3 sec		
Spectral Noise(1 Hz)	800 µg/√Hz	7840 (µm/sec²)/√Hz	[1]	
Spectral Noise(10 Hz)	250 µg/√Hz	2450 (µm/sec²)/√Hz	[1]	
Spectral Noise(100 Hz)	60 µg/√Hz	590 (µm/sec²)/√Hz	[1]	
Spectral Noise(1 kHz)	50 µg/√Hz	490 (µm/sec²)/√Hz	[1]	
Spectral Noise(10 kHz)	40 µg/√Hz	392 (µm/sec²)/√Hz	[1]	
Electrical Isolation(Base)	>10⁸ Ohm	>10⁸ Ohm		
Physical				
Size (Height x Length x Width)	0.14 in x 0.45 in x 0.25 in	3.6 mm x 11.4 mm x 6.4 mm		
Weight	0.017 oz	0.5 gm	[1]	
Sensing Element	Ceramic	Ceramic		
Sensing Geometry	Shear	Shear		
Housing Material	Anodized Aluminum	Anodized Aluminum		
Sealing	Epoxy	Epoxy		
Electrical Connector	3-56 Coaxial Jack	3-56 Coaxial Jack		
Electrical Connection Position	Side	Side		
Mounting	Adhesive	Adhesive		

OPTIONAL VERSIONS

Optional versions have identical specifications and accessories as listed for the standard model except where noted below. More than one option may be used.

NOTES:

[1] Typical.
[2] Zero-based, least-squares, straight line method.
[3] See PCB Declaration of Conformance PS023 for details.

SUPPLIED ACCESSORIES:

Model 030A10 Coax Cable, 10 ft (3 m), 3-56 plug to 10-32 plug. (1)
Model 039A27 One-piece removal tool for Models 352C22, 357C10, 352A21, & 357A09 (1)
Model 080A109 Petro Wax (1)
Model ACS-1 NIST traceable frequency response (10 Hz to upper 5% point). (1)

Entered: AP	Engineer: JJB	Sales: WDC	Approved: JJB	Spec Number:
Date: 12/16/2013	Date: 12/16/2013	Date: 12/16/2013	Date: 12/16/2013	10668

All specifications are at room temperature unless otherwise specified.
In the interest of constant product improvement, we reserve the right to change specifications without notice.
ICP® is a registered trademark of PCB Group, Inc.

PCB PIEZOTRONICS
3425 Walden Avenue, Depew, NY 14043
Phone: 716-684-0001
Fax: 716-684-0987
E-Mail: info@pcb.com

Fig. 2.5 Accelerometer manufacturer's specifications

The random component of the overall uncertainty value (row 12) was calculated from test data using statistical methods. When analyzing test data, it is preferable to have pairs of instrumentation collecting duplicate data. This allows a standard deviation to be calculated based on the variation of the data, and therefore gives a more accurate random error component to be used in the overall uncertainty calculation. A search for available test data from RQVV test programs did not produce such a setup. Therefore, a simple test setup was assembled that would give the desired data. Figure 2.6 shows the setup. A rectangular aluminum plate (approximately 10 × 3 × 0.05 inches) was mounted to the 4000 lbf shaker. One pair of accelerometers was mounted side-by-side approximately one inch from the end on the plate centerline. The test spectrum was a random input that was controlled from 0 to 2000 Hz. The three tests were run at different Grms levels: 3, 4, and 5. Figure 2.7 is an acceleration vs. frequency plot for the 3 Grms test. The plots for the 4 and 5 Grms tests are similar. Figure 2.8 shows a linearity plot for the 3 Grms test. The linearity plots for the 4 and 5 Grms tests are similar. Note that the slope of the line (Fig. 2.8) is not unity as would be expected. The likely cause is the measurement error of the accelerometers.

The random error entry in row 12 was determined using the following steps.

1. The equation of the line for each linearity plot was determined.
2. The corresponding theoretical Y axis value was calculated for each X axis input.
3. The difference (delta) between the theoretical and measured Y axis values was calculated.
4. The standard estimate of error (SEE) of the deltas was calculated for each of the three test runs and used as an estimate of the standard deviation of the measurements once divided by the square root of two.
5. The three standard deviations were pooled to get the random error component. This value (0.32) was entered in the spreadsheet.

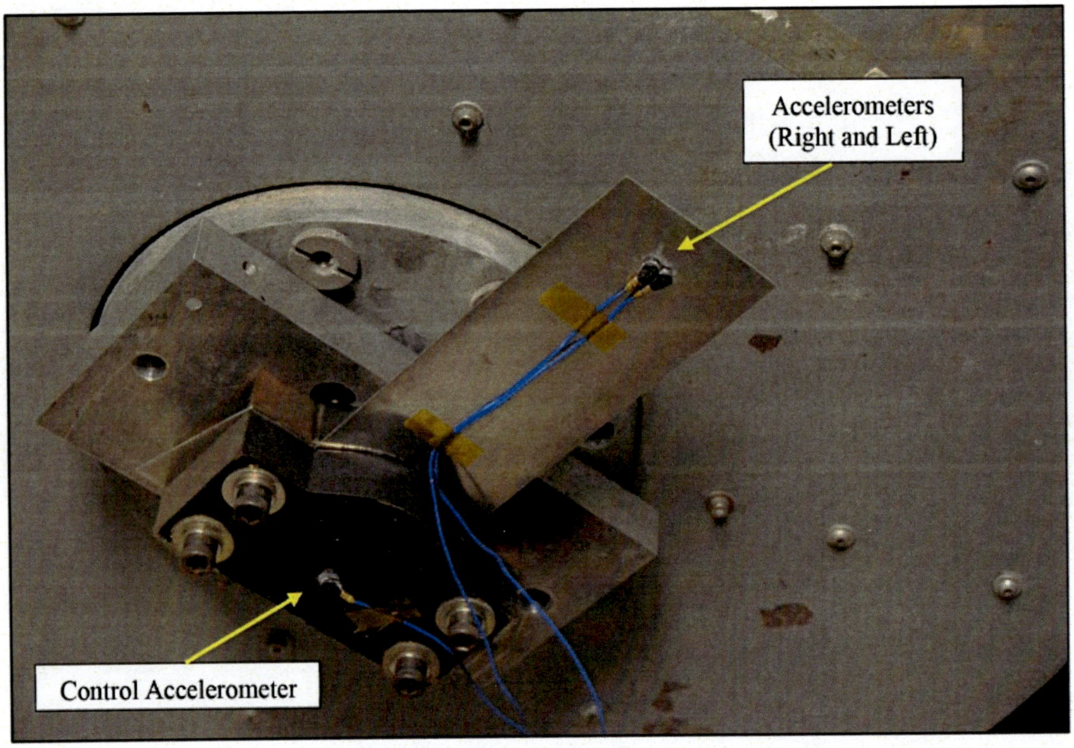

Fig. 2.6 Test setup

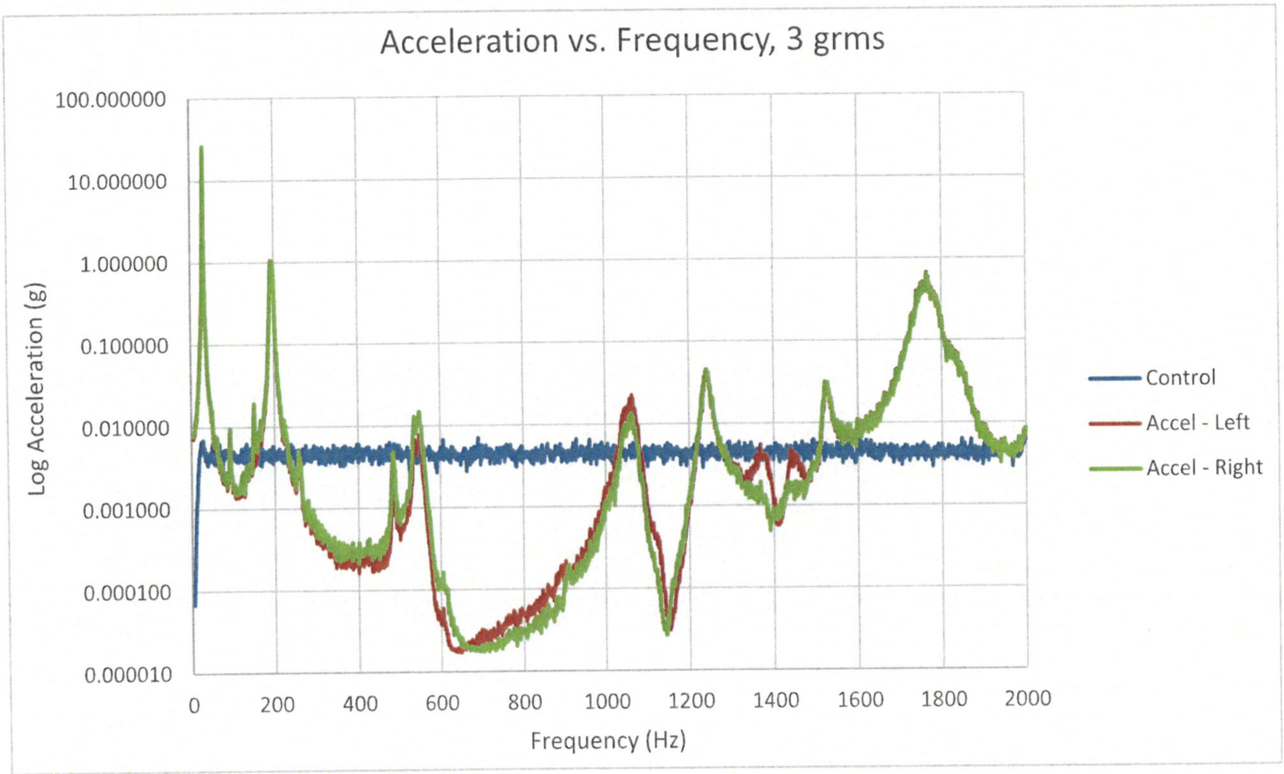

Fig. 2.7 Acceleration vs. frequency plot, 3 Grms

The random component ultimately had little impact on the final uncertainty value. It contributed 0.2% of the combined uncertainty value.

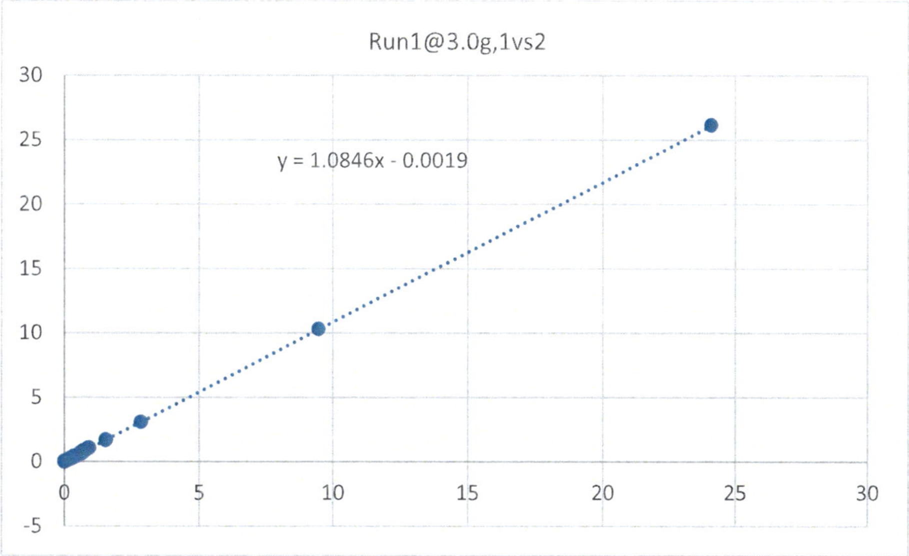

Fig. 2.8 Linearity plot, 3 Grms

2.11 Conclusion

The measurement uncertainty program that was completed by AFRL/RQVV was intended to provide "user friendly" procedures and tools to calculate and document test data measurement uncertainty values. AFRL/RQVV customers will benefit by understanding the quality of the test data. While the project consultant (Mr. Ron Dieck) performed the bulk of the initial uncertainty calculations, this project also provided a wealth of experience and information to RQVV personnel on the many facets of the uncertainty calculation process. This includes an understanding of which systematic error components can be regarded as negligible and which need to be closely examined. Random error components cannot always be easily calculated from available test data. Test data must follow a normal distribution to be useful. However, in some situations, the random error component is also insignificant when compared to the systematic errors.

Reference

1. Dieck, R.H.: Measurement Uncertainty, Methods and Applications, 5th edn. ISA, Research Triangle Park (2017)

Chapter 3
Carbon Nanotube (CNT) Elastomers for Sensing Applications: A Narrative Review

Hannah Loukusa and Eric Little

Abstract Elastomers are traditionally considered materials with low tensile strength and minimal electrical conductivity. Carbon nanotubes (CNT) are classified as fullerenes, or carbon-based materials forming a molecular mesh, that assume a cylindrical shape and display remarkable properties, such as high tensile strength and enhanced electrical conductivity as compared to other carbon allotropes. Elastomers doped with carbon nanotubes form CNT elastomers, which act as semi-conductors and can be formulated to exhibit many useful characteristics for the design of various sensors, such as strain gages. While CNT elastomers have been studied for more than 20 years, there are relatively few, if any, commercial sensor products utilizing this innovative technology. In this chapter, we survey selected portions of the CNT literature, explore the operation and characteristics of CNT elastomers, and consider some CNT elastomer requirements of potential interest for the design of sensors.

Keywords Carbon nanotubes · CNT · Elastomer · Strain gauge · Strain gage

3.1 Polymers

Polymers are a generic term for any molecule, organic or synthetic, composed of repeating subunits. Formed by a polymerization process, or the chemical reactions used to join constituent monomers into a polymer chain, polymers play important roles in modern daily life. For example, plastics are synthetic (or sometimes semi-organic) polymers composed of hydrocarbon monomers [1]. Even deoxyribonucleic acids (or DNA) that forms the basis of human life are polymers comprised of nucleotide monomers [2]. In their unprocessed or raw state, polymers often assume the shape of long, flexible strings, much like a cooked spaghetti noodle [3], and are not particularly useful in the construction of practical products. Further processing of raw polymers, such as by introducing small amounts of selected additives, can change the bulk characteristics of those polymers into useful engineering materials.

3.2 Elastomers

Elastomers are a class of polymers characterized by occasional cross-linking of the long-chain polymers, such as by covalent cross-linking [4]. The stiffness of an elastomer is governed by the number of polymer cross-links, such as increasing cross-link density results in stiffer elastomers [5]. The cross-linking process can include the addition of a cross-link promoter (or "cross-linker") to the raw polymer and then exposing the combined materials to either a source of energy, such as gamma radiation, or chemical activation, such as by mixing the combined materials with peroxide [6]. The elasticity of an elastomeric material is characterized by its ability to "move" or otherwise reconfigure its amorphous structure under an applied load, yet return to its original structural configuration after the load is removed due to the "memory" of the material introduced by the cross-linking process [7].

H. Loukusa · E. Little (✉)
Mechanical & Manufacturing Engineering, St. Cloud State University, St. Cloud, MN, USA
e-mail: hlloukusa@go.stcloudstate.edu; eflittle@stcloudstate.edu

C. Walber et al. (eds.), *Sensors & Instrumentation and Aircraft/Aerospace Testing Techniques, Volume 8*, Conference Proceedings of the Society for Experimental Mechanics Series, https://doi.org/10.1007/978-3-031-34938-6_3

3.3 Compounding

The process of compounding can further refine the properties of an elastomer through the addition of fillers or additives that can impart special characteristics to the elastomer, such as color, impermeability, or oil resistance [8]. Fillers can also be added to "reinforce" an elastomer, such as to improve tensile strength, abrasion resistance, and extension moduli. For example, carbon black, a paracrystalline form of carbon with a high surface-area-to-volume ratio [9], is a common elastomer reinforcement filler [10]. Owing to surface chemistry, where a large percentage of carbon atoms are available to interact with the polymer matrix, carbon black is hypothesized to increase elastomer tensile strength through physical adsorption with the elastomeric polymer and covalent bonding at cross-linking points [11]. Another example of a carbon-based elastomer filler is the carbon nanotube (or CNT).

3.4 Carbon Nanotubes (CNT)

Originally synthesized and identified in 1952 [12] and subsequently "rediscovered" in 1991 [13], the CNT is a fullerene or a carbon-based material forming a molecular mesh that assumes a cylindrical shape [14]. See Fig. 3.1.

Physically, the diameter of a CNT can vary in the range of about 0.4–40 nanometers (nm) with lengths from 0.14 nm to 55.5 cm [16]. See Fig. 3.2. CNT display unusual tensile strength (50+ GPa) as compared to carbon steel (540 MPa) [17]. Single-walled CNT are predicted to have an electrical resistance of $\sim$6 kΩ independent of length and diameter [17].

3.5 CNT Elastomers

Apart from their enhanced mechanical properties, elastomers doped with CNT fillers also exhibit advantageous electrical properties. While elastomers are typically considered an insulating material unable to conduct electrical charge, CNT-doped elastomers act as conducting materials, capable of transmitting a current through the elastomer in response to an applied

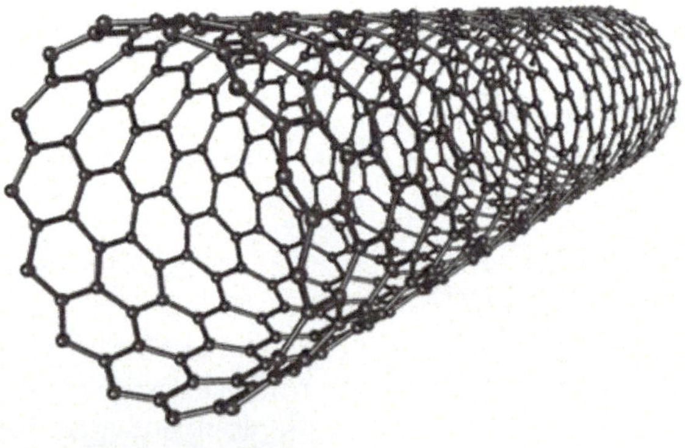

Fig. 3.1 Molecular structure of a single-walled carbon nanotube (SWCNT) [15]

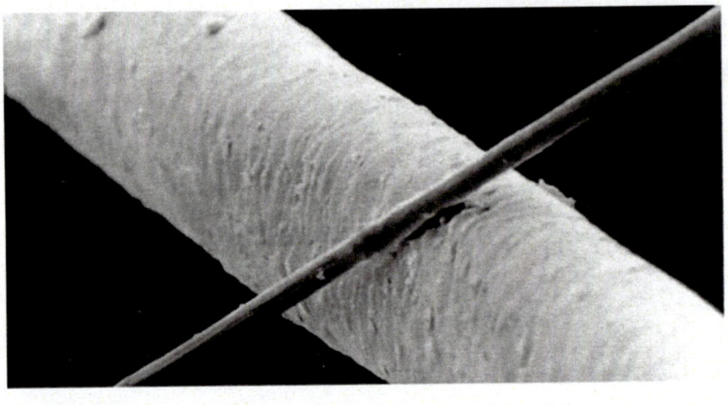

Fig. 3.2 Scale of a CNT (thin tube) compared to a human hair (thick tube; diameter = $\sim$50 nm) [18]

voltage [19]. Electrical conductivity is induced as the CNT filler is dispersed throughout the insulating elastomer at a density sufficient to create a conducting network through the elastomer, such as by direct contact of CNTs creating a conduction path or by CNTs in close proximity to each other in a process known as "tunneling" [20]. In an example, CNT elastomers can be formed into the shape of a sensor, such as a resistance strain gauge. As load is applied to the CNT elastomer strain gauge, the elastomer material deforms, interrupting current transmission paths by "breaking" connections between conducting CNT in the elastomer matrix, increasing the electrical resistance of the strain gauge [21, 22]. As the load is released, the elastic nature of the CNT elastomer will cause the gauge to return to its undeformed state, reestablishing the electrical CNT connections and reducing electrical resistance of the strain gauge.

3.6 Factors Affecting CNT Elastomer Conductivity

Figures 3.3 and 3.4 illustrate the effect of CNT concentration on conductivity within an elastomer matrix [23]. In illustration (a) of Fig. 3.3, a low concentration of well-dispersed CNT does not allow for contact between CNT particles, resulting in low electrical conductivity shown in Fig. 3.4 at location a. In illustration (b) of Fig. 3.3, increased CNT concentration brings CNT particles into closer proximity and promotes "tunneling" effects [21], where electrical current can pass without physical connection between CNT particles and increases conductivity as shown in Fig. 3.4 at location b. In illustration (c) of Fig. 3.3, increased CNT concentration allows CNT particles to contact one another and provides a continuous electrical pathway through the elastomer (shown as red CNT particles in illustration (c)) to achieve the so-called percolation threshold shown in Fig. 3.4 at location c. Note that the percolation threshold can be considered the CNT concentration that will lead to an abrupt increase in electrical conductivity (or correspondingly, a sharp decrease in electrical resistance) through the elastomer. In illustration (d) of Fig. 3.3, further increases in CNT concentration lead to multiple flanking electrical paths (shown as blue CNT particles in illustration (d)) and ultimately, electrical saturation of the elastomer shown in Fig. 3.4 at location d.

In addition to CNT concentration, volumetric distribution of the CNT filler in the elastomer matrix has been reported to affect conductivity [24]. Due at least in part to surface chemistry, CNT particles tend to agglomerate prior to elastomeric compounding to form "balls" or clumps of CNT [25] thought to be drawn together by ionic bonding. In general, compounding techniques that generate a more even dispersion of CNT particles through the elastomer matrix will result in a larger number of conduction paths to increase CNT elastomer conductivity and potentially lower the amount of CNT filler required to achieve the percolation threshold.

The use of CNT elastomers in sensor applications have been thoroughly explored in the literature. For example, reference to pressure sensing applications include work on "smart" skin, such as for haptic sensing applications [26], and sandwich-type piezoresistive pressure sensors [27], such as for measurement of mechanical loads. Similarly, considerable effort has been directed to strain sensing applications including sensors for human motion detection [28–30], roadway assessment [31–33], and general strain measurement purposes [34–36]. Many of these studies report excellent results, including substantially

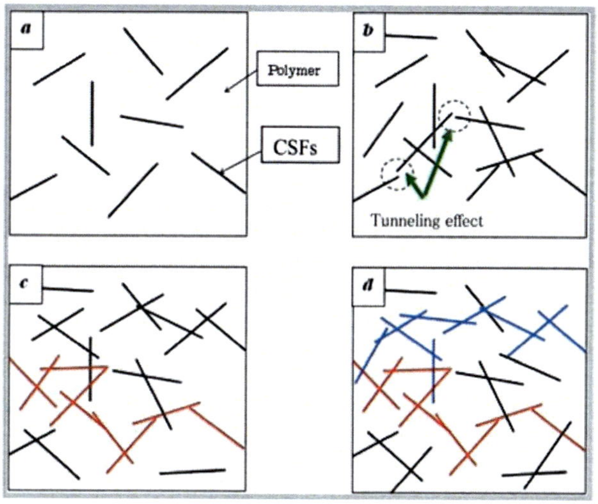

Fig. 3.3 (left) CNT concentration in the elastomer matrix can affect conductivity [23]

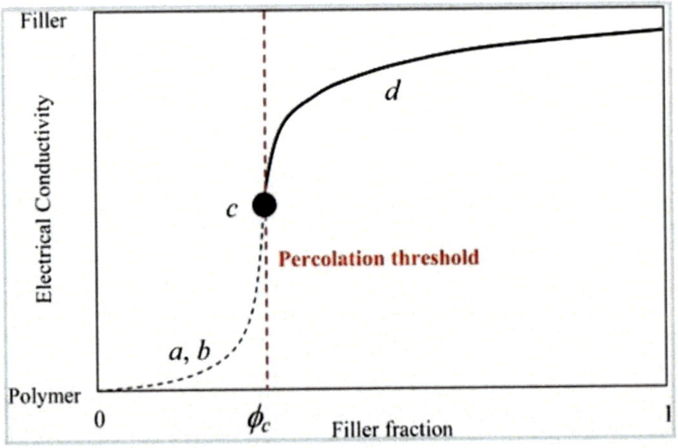

Fig. 3.4 (right) A graph of filler fraction to electrical conductivity [23]

linear response and little to no hysteresis characteristics [37]. However, even with the benefit of more than 30 years of academic research and discussion, the authors are unable to find any commercial products utilizing CNT elastomers.

At least two issues have been suggested to explain the lack of widespread CNT elastomer adoption – stability and repeatability [21]. Stability refers to fluctuations in sensor resistance not associated with mechanical loading whereas repeatability is characterized as the "repeatable resistance-strain performance" of a sensor [21]. Similar to their foil strain gauge counterparts, CNT strain gauges can experience environmental factors that affect the stability of resistance measurements, such as temperature sensitivity [38] and humidity [39], as well as factors unique to CNT elastomers, such as chemical effects [40] and particle size effects [41]. One source of undesirable resistance repeatability is manifested with the presence of hysteresis [42], which has been suggested to indicate the degradation of the CNT elastomer while in use. It is hypothesized the as least some of the challenges related to stability and repeatability can be addressed with improved CNT elastomer processing.

3.7 Factors Affecting CNT Elastomer Stability and Repeatability

There are several factors that can be controlled, or otherwise influenced, at the elastomer compounding stage that may improve the stability and repeatability of CNT elastomer response.

3.7.1 Polymer Matrix

Proper selection of the polymer matrix optimized for a particular characteristic might reduce stability challenges associated with CNT elastomers. Viscosity of the polymer matrix might be selected based upon the type of compounding force used [23]. For example, relatively inviscid polymers can be selected where a high-speed shear mixer is intended for compounding, whereas relatively viscous polymers might be better suited for sonication dispersion. Similarly, selection of a polymer with an appropriate molecular weight can be considered to facilitate CNT tunneling in polymers with low weight-faction CNT loads. Further, compounding fillers can be selected to control moisture content [41], which has been shown to cause considerable variation in the measured resistance of unstrained structural cement applications [43].

3.7.2 CNT Particles

Several aspects of CNT particles can affect the stability and repeatability characteristics of CNT elastomers. The size, shape, and aspect ratio of CNT particles have been noted as variables that can affect the measured resistance of a CNT elastomer [23]. Molecular functionalization, or the process of adding new capabilities or properties to a material by changing the surface chemistry of the material, can be applied to CNT, such as to improve CNT dispersion in an elastomer matrix [13, 44] or improve conductivity [45]. Further, the type of CNT used, such as single-walled carbon nanotube (SWCNT), multi-walled carbon nanotube (MWCNT), or thin-walled carbon nanotubes (TWCNT), can affect CNT elastomer performance. For example, several studies have suggested CNT elastomers can demonstrate piezo-electric characteristics with the use of MWCNT particles compounded into an elastomer matrix [46, 47].

3.7.3 Compounding Variables

The methods of CNT dispersion can also play a significant role in the stability and repeatability of CNT elastomers. CNT loading of an elastomer, or the amount of CNT particles required to achieve a percolation threshold, can adjust the conductivity of a CNT elastomer. For example, increasing CNT loading can potentially improve overall repeatability by increasing the volume fraction of CNT in contact with the elastomer matrix. However, additional CNT loading will also increase bulk stiffness of the elastomer, such as to reduce sensitivity of the CNT elastomer to displacement. The mechanical compounding process can also affect stability and repeatability. For example, two common compounding techniques, shear mixing and sonication, induce CNT dispersion in substantially different ways. The effects of mixing speed, mixing time, and other variables have been reported to impact the electrical properties of the CNT elastomer [23]. Curiously, curing temperature and mixing conditions have also been identified as significant variables influencing the formation of the CNT conducting network [48], which could affect the stability and repeatability of CNT elastomer materials.

3.8 Conclusions

CNT elastomers are materials with interesting electrical properties that offer many opportunities for sensor applications. However, despite extensive research and a long history of publications, few (if any) commercial products utilize CNT elastomers as a component in their construction. Two characteristics of CNT elastomers, the stability of electrical resistance response and the repeatability of electrical resistance response in elastomers subjected to dynamic loading, have been identified as potentially limiting factors to the use of CNT elastomers in sensor applications. It is hypothesized that improved control of the CNT elastomer compounding process could play a role in increasing the stability and repeatability of CNT elastomer material response.

References

1. https://omicoplastics.com/blog/plastics-polymers-difference/. Accessed Aug 2022
2. https://www.ebi.ac.uk/training/online/courses/biomacromolecular-structures/dna/. Accessed Aug 2022
3. https://www.cmu.edu/gelfand/lgc-educational-media/polymers/what-is-polymer/index.html#:~:text=Polymers%20have%20interesting%20behavior%20and,in%20a%20bowl%20of%20pasta. Accessed Aug 2022
4. https://www.osborneindustries.com/news/elastomers-vs-polymers/#:~:text=An%20elastomer%20is%20made%20from,elongation%2C%20flexibility%2C%20and%20elasticity. Accessed Aug 2022
5. https://en.wikipedia.org/wiki/Cross-link. Accessed Aug 2022
6. https://www.mddionline.com/news/cross-linking-thermoplastic-elastomers-improved-product-performance. Accessed Aug 2022
7. Özdemir, T.: Chapter 5: Elastomeric micro- and nanocomposites for neutron shielding. In: Micro and nanostructured composite materials for neutron shielding applications, pp. 125–137 (2020). https://doi.org/10.1016/B978-0-12-819459-1.00005-2
8. Pajarito, B.: Differentiation of non-black fillers in rubber composites using linear discriminant analysis of principal components. J. Sci. Eng. Compos. Mater. **26**, 282–291 (2019)
9. https://en.wikipedia.org/wiki/Carbon_black. Accessed Sept 2022
10. Rothon, R.: Particulate fillers in elastomers. In: Polymers and Polymeric Composites: A Reference Series Book Series, pp. 125–146 (2017)
11. Robertson, C., et al.: Nature of carbon black reinforcement of rubber. Perspect. Orig. Polym. Nanocompos. **13**(4), 538 (2021)
12. Katz, E.: Credit to Pioneering Work on Carbon Nanotubes. https://arxiv.org/ftp/arxiv/papers/2105/2105.03658.pdf. Accessed Oct 2022
13. Wong, K.-K., et al.: Mechanical and thermal behavior of a polymer composite reinforced with functionalized carbon nanotubes. Key Eng. Mater. **334–335**, 705–708 (2007)
14. Gomez-Bualdron, D., et al.: Chapter 5: Carbon nanotubes: engineering biomedical applications. In: Progress in Molecular Biology and Translational Science, vol. 104, pp. 175–245 (2011). https://doi.org/10.1016/b978-0-12-416020-0.00005-x
15. https://innovationsgesellschaft.ch/en/evaluation-of-the-health-effects-of-carbon-nanotubes/. Accessed July 2022
16. https://en.wikipedia.org/wiki/Carbon_nanotube. Accessed Aug 2022

17. Kang, I., et al.: Introduction to carbon nanotube and nanofiber smart materials. Compos. Part B. **37**(6), 382–394 (2006)
18. https://www.reddit.com/r/pics/comments/5whh0p/a_single_carbon_nanotube_on_a_human_hair_for_scale/. Accessed Aug 2022
19. Mensah, B., et al.: Carbon nanotube-reinforced elastomeric nanocomposites: a review. Int. J. Smart Nano Mater. **6**(4), 211–238 (2015)
20. Prabhakar, R., et al.: Tunneling-limited thermoelectric transport in carbon nanotube networks embedded in poly(dimethylsiloxane) elastomer. ACS Appl. Energy Mater. **2**(4), 2419–2426 (2019)
21. Obitayo, W., et al.: A review: carbon nanotube-based Piezoresistive strain sensors. J. Sens. (2012). https://doi.org/10.1155/2012/652438
22. Kim, T.G., et al.: Electrically conductive silicone-based nanocomposites incorporated with carbon nanotubes and silver nanowires for stretchable electrodes. ACS Omega. **6**(47), 31876–31890 (2021)
23. Alamusi, et al.: Piezoresistive strain sensors made from carbon nanotubes based polymer nanocomposites. Sensors (Basel). **11**(11), 10691–10723 (2011)
24. Liu, C.-X., et al.: Improved dispersion of carbon nanotubes in polymers at high concentrations. Nanomaterials (Basel). **2**(4), 329–347 (2012)
25. Hnog, J.S.: Extension-induced dispersion of multi-walled carbon nanotube in non-Newtonian fluid. J. Rheol. **51**, 5 (2007)
26. He, Y., et al.: A breathable, sensitive and wearable piezoresistive sensor based on hierarchical micro-porous PU@CNT films for long-term health monitoring. Compos. Sci. Technol. **200**, 108419 (2020)
27. Ali, A.: Pressure sensitive sensors based on carbon nanotubes, graphene, and its composites. J. Nanomater. **2018** (2018). https://doi.org/10.1155/2018/9592610
28. Demidenko, N., et al.: Flexible strain-sensitive silicone-CNT sensor for human motion detection. Bioengineering (Basel). **9**(1), 36 (2022)
29. Shu, W.-B., et al.: Direct ink writing of graphene/Cnt/silicone composite strain sensor with near-zero temperature coefficient of resistance. J. Mater. Chem. C. (2022). https://doi.org/10.2139/ssrn.4022067
30. Lu, Y., et al.: Recent developments in bio-monitoring via advanced polymer nanocomposite-based wearable strain sensors. Biosens. Bioelectron. **123**, 167–177 (2019)
31. Xin, X., et al.: The innovative self-sensing strain sensor for asphalt pavement structure: substitutability and synergy effects of graphene platelets with carbon nanotubes in epoxy composites. Front. Mater. (2022). https://doi.org/10.3389/fmats.2022.824364
32. Yu, X., et al.: Carbon Nanotube Based Self-Sensing Concrete for Pavement Structural Health Monitoring. https://rosap.ntl.bts.gov/view/dot/40249
33. Camacho-Ballesta, C., et al.: Performance of cement-based sensors with CNT for strain sensing. Adv. Cem. Res. **28**(4), 1–11 (2015)
34. Thompson, B., et al.: Aerosol printed carbon nanotube strain sensor. Proc. SPIE Int. Soc. Opt. Eng. (2012). https://doi.org/10.1117/12.914964
35. Lee, D., et al.: Microfabrication and characterization of spray-coated single-wall carbon nanotube film strain gauges. Nanotechnology. **22** (2011). https://doi.org/10.1088/0957-4484/22/45/455301
36. Kang, I., et al.: Introduction to carbon nanotube and nanofiber smart materials. Compos. Part B Eng. **37**(6), 382–394 (2006)
37. Cai, L., et al.: Super-stretchable, transparent carbon nanotube-based capacitive strain sensors for human motion detection. Sci. Rep. (2013). https://doi.org/10.1038/srep03048
38. Jang, S.-H., et al.: Carbon nanotube-reinforced smart composites for sensing freezing temperature and deicing by self-heating. Nanomater. Nanotechnol. (2018). https://doi.org/10.1177/1847980418776473
39. Saleem, M., et al.: Humidity sensing properties of CNT–OD–VETP nanocomposite films. Physica E. **43**(1), 28–32 (2010)
40. Li, X., et al.: Multiwalled carbon nanotube film for strain sensing. Nanotechnology. **19**(4), 045501 (2008)
41. Loh, K., et al.: Multifunctional layer-by-layer carbon nanotube–polyelectrolyte thin films for strain and corrosion sensing. Smart Mater. Struct. **16**(2), 429 (2007)
42. DeFocatiis, D., et al.: Electromechanical hysteresis in filled elastomers. In: Constitutive Models for Rubber VII, Proceedings of the 7th European Conference on Constitutive Models for Rubber. Taylor & Francis (2011). https://doi.org/10.1201/b11687-45
43. Teng, F., et al.: Piezoresistive/piezoelectric intrinsic sensing properties of carbon nanotube cement-based smart composite and its electromechanical sensing mechanisms: a review. Nanotechnol. Rev. (2021). https://doi.org/10.1515/ntrev-2021-0112
44. Dubey, R., et al.: Functionalized carbon nanotubes: synthesis, properties and applications in water purification, drug delivery, and material and biomedical sciences. Nanoscale Adv. (2021). https://doi.org/10.1039/D1NA00293G
45. Jeon, I.-Y., et al.: Functionalization of carbon nanotubes. In: Carbon Nanotubes – Polymer Nanocomposites (2011). https://doi.org/10.5772/18396
46. Il'ina, M., et al.: Piezoelectric response of multi-walled carbon nanotubes. Materials. (2018). https://doi.org/10.3390/ma11040638
47. Il'ina, M., et al.: Analysis of the piezoelectric properties of aligned multi-walled carbon nanotubes. Nanomaterials (Basel). (2021). https://doi.org/10.3390/nano11112912
48. Hu, N., et al.: Effect of fabrication process on electrical properties of polymer/multi-wall carbon nanotube nanocomposites. Compos. A: Appl. Sci. Manuf. **39**(5), 893–903 (2008)

Chapter 4
Magnetoelastic Vibration Sensors

Ehsan Vatankhah, Connor Hodges, Xiaoyu Niu, Zihuan Liu, Yuqi Meng, Askold S. Belyakov, and Neal A. Hall

Abstract In this work, we explore the application of magnetoelasticity towards the realization of vibration transducers. Such sensors have unique and complementary properties when compared against more conventional transduction schemes. The sensors can be constructed to have a native response to jerk, the time derivative of acceleration. Acceleration causes force on an internal mass, the force induces strain on a magnetoelastic material, and the strain produces a magnetic field. The time derivative of magnetic flux through a voice coil results in a measurable voltage, consistent with Faraday's law. Taken together, the generated voltage is directly proportional to applied jerk. Because of this native jerk response, the transducer's frequency response has a 6 dB/octave slope compared to accelerometers and a 12 dB/octave slope compared to geophones. A second noteworthy feature is that, like geophones, magnetoelastic sensors have low output impedance and can, therefore, be used without amplification electronics in many applications. This may be ideal for applications requiring large networks of sensors.

Keywords Moving coil geophone · Sensitivity · Noise · Piezomagnetic

4.1 Introduction

Magnetostriction refers to a material's ability to produce strain in response to an externally applied magnetic field. The inverse effect, known as magnetoelasticity, or *Villari* effect, refers to a material's ability to produce or alter a magnetic field in response to mechanical stress. When appropriately biased with a magnetic field and/or a mechanical prestress, magnetostrictive materials exhibit magnetoelasticity. Much like piezoelectric materials, there are many naturally occurring materials that exhibit magnetostriction and many more alloys that have been engineered specifically to enhance magnetostriction. Some examples of magnetostrictive materials include Ni (Nickel); Fe (Iron); alloys such as Al-Fe (Alfer), Ni-Fe (Permalloy), Co-Ni, Co-Fe, and Co-Fe-Ni (Permendur and Perminvar); and Cobalt and Nickel ferrites such as $CoFe_2O_4$ and $NiFe_2O_4$ [1]. In more recent years, engineered materials such as Terfenol-D (Tb-Dy-Fe) and Galfenol (Ga-Fe) have been developed to have a larger magnetostrictive coefficient and higher magneto-mechanical coupling coefficient [2]. These materials have been used as actuators for their large displacement, but their potential for sensing has been less explored.

4.2 Background

Magnetoelasticity can be exploited to create pressure, force, and inertial sensors. When a material is biased, for example, with a static external field of magnitude H, and undergoes a dynamic strain as a function of time $s(t)$, a time-varying magnetic flux density $B(t)$ results that can be sensed. Precisely these physics are exploited in the proposed magnetoelastic geophone, depicted in Fig. 4.1. Permanent magnets establish the required static or biasing field within the magnetoelastic material,

E. Vatankhah (✉) · C. Hodges · X. Niu · Z. Liu · Y. Meng · N. A. Hall
Chandra Department of Electrical and Computer Engineering, University of Texas at Austin, Austin, TX, USA
e-mail: e.vatankhah@utexas.edu; cmhodges@utexas.edu; xyniu@utexas.edu; zihuanliu@utexas.edu; yuqimeng@utexas.edu; nahall@mail.utexas.edu

A. S. Belyakov
Retired Geophysicist, San Francisco, CA, USA

C. Walber et al. (eds.), *Sensors & Instrumentation and Aircraft/Aerospace Testing Techniques, Volume 8*, Conference Proceedings of the Society for Experimental Mechanics Series, https://doi.org/10.1007/978-3-031-34938-6_4

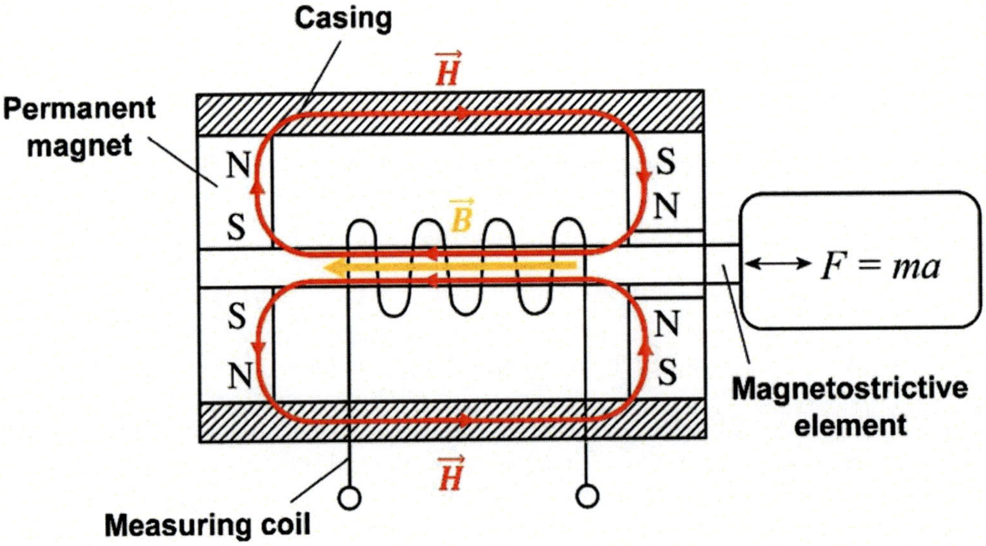

Fig. 4.1 Schematic of the magnetoelastic sensor

as shown in the figure. Externally applied dynamic force $F(t)$ gives rise to strain $s(t)$ in the magnetoelastic material, in turn producing a field $B(t)$ directed along the axis of the magnetoelastic material. The time-varying field $B(t)$ induces an electromotive force (i.e., a measurable voltage) in a coil wrapped around the magnetoelastic material. Consistent with Faraday's Law of Induction, the measured voltage across the coil winding is proportional to dB/dt, and, therefore, to dF/dt, the time derivative of the externally applied force.

In acceleration-sensing applications, F results from acceleration of a proof mass attached to the magnetoelastic element. The coil voltage is, therefore, proportional to the time derivative of acceleration or, equivalently, the second time derivative of velocity. This contrasts with conventional geophones, which respond in proportion to surface or ground velocity. Stated in frequency domain terms, conventional geophones are said to be flat with respect to velocity, while the magnetoelastic sensor in Fig. 4.1 is expected to have a response that increases at +12 dB/octave with respect to ground velocity.

In other forms, the magnetoelastic effect has been used in wireless noncontact sensors for torque and pressure. The class of wireless magnetoelastic resonance sensors (where an excitation coil and a sense coil reside away from the magnetoelastic material undergoing strain) has been actively explored in recent history. Reference [3] from 2019 is a review article citing more than 75 articles.

4.3 Measurement and Analysis

A prototype sensor based on the schematic in Fig. 4.1 and as described in reference [4] has been fabricated. Basic characterization has begun in our lab at UT Austin. The magnetoelastic element is a cylindrical shell with inner diameter equal to 21.1 mm, outer diameter equal to 32.1 mm and 58.0 mm in length. The cylinder is iron alloyed with silicon and aluminum. The prototype was tested alongside a commercial 10 Hz geophone (model RTC 10 375 with 10 Hz suspension resonance frequency and 375 Ω coil resistance [5]). In our acoustics laboratory at UT Austin, the two devices were placed in close proximity, and both recorded data into a 24-bit data logger. Ambient background vibration on the lab floor served as a broadband stimulus into both devices. The magnitude of the magnetoelastic sensor's FFT was normalized to the magnitude of the commercial geophone's FFT. Time segments of 12.8 s were used. Twenty such time segments were captured, the FFTs were taken and normalized, and the average of all 20 normalizations were saved. The result is presented in Fig. 4.2. The response is smooth in the 10–200 Hz range. Outside of this frequency range, the commercial geophone used for normalization does not have a reliable response due to spurious resonances and/or poor sensitivity. In the reliable 10–200 Hz characterization range, the magnetoelastic sensor's response is observed to increase at 12 dB/octave relative to that of the commercial geophone, just as expected. Since the commercial geophone is flat with respect to velocity in this frequency range, this implies that the magnetoelastic prototype responds to jerk, i.e., the time derivative of acceleration.

Because the RTC geophone is calibrated and has a known flat response to velocity with a known calibration constant, we can quantify the magnetoelastic sensor's sensitivity over this frequency range via this normalization. For example, the 0 dB

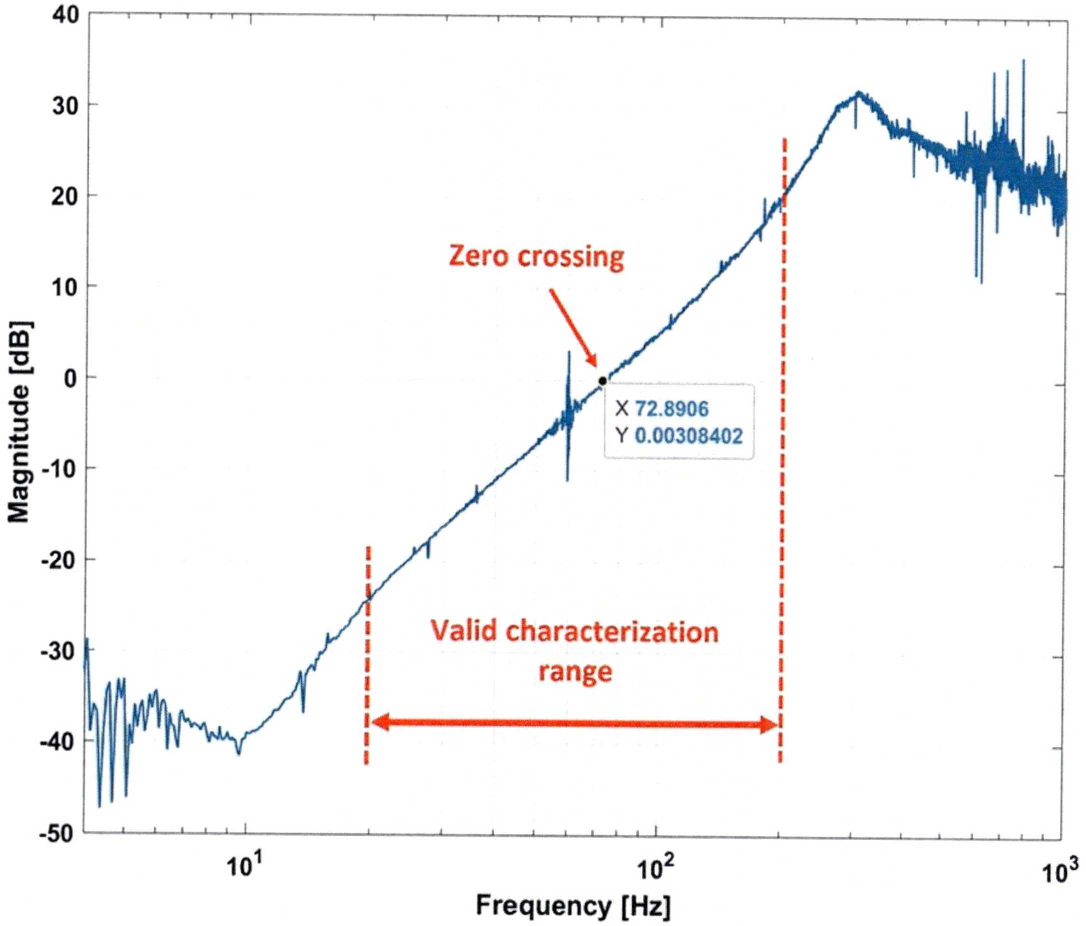

Fig. 4.2 Magnetoelastic sensor response normalized to the response of a commercial geophone. The shaded region indicates the valid characterization range. At zero dB, the sensitivity of the magnetoelastic sensor equals that of the moving coil geophone

crossing in the figure is at 73 Hz. The prototype sensor has the same sensitivity as the geophone at this frequency, which is known via the data sheet to be 28.8 V/[m/s].

The commercial geophone and the magnetoelastic prototype can be further explored by comparing their measured electrical impedance. Impedance spectroscopy is an established technique for many sensors and actuators including loudspeakers, geophones, and piezoelectric devices [6]. Figure 4.3a plots the real part of the measured impedance (i.e., the resistance), while Fig. 4.3b plots the imaginary component of the impedance (i.e., the reactance). The coil resistance is the real part of the measured impedance measured at low frequencies and off-resonance. Figure 4.3a indicates that the commercial RTC geophone has a coil resistance equal to 375 Ω and a suspension resonance equal to 10 Hz, as expected. Further inspection of Fig. 4.3a reveals a coil resistance equal to 3.9 kΩ and a fundamental resonance equal to approximately 2 kHz for the magnetoelastic sensor. The magnetoelastic prototype does not employ springs, so this fundamental resonance is a structural resonance of the rod-mass system depicted in Fig. 4.1.

Signal-to-noise ratio (SNR) is one of many figures of merit for sensors and is often important to consider. Although no direct SNR assessment is presented herein, one can compare simulated or predicted self-noise of each sensor based on insights from the measurements above. The electrical self-noise voltage at each sensor output is computed as

$$\tilde{v}_n = \sqrt{4k_B T Re\{Z_{\text{in}}\}} \left[\frac{V}{\sqrt{Hz}}\right] \tag{4.1}$$

where the real part of Z_{in}, $Re\{Z_{\text{in}}\}$ is obtained from measurements presented in Fig. 4.3. The sensitivities of each sensor are known from data presented in Fig. 4.2. The input-referred noise is synonymous with the minimum detectable signal (MDS) and is obtained as

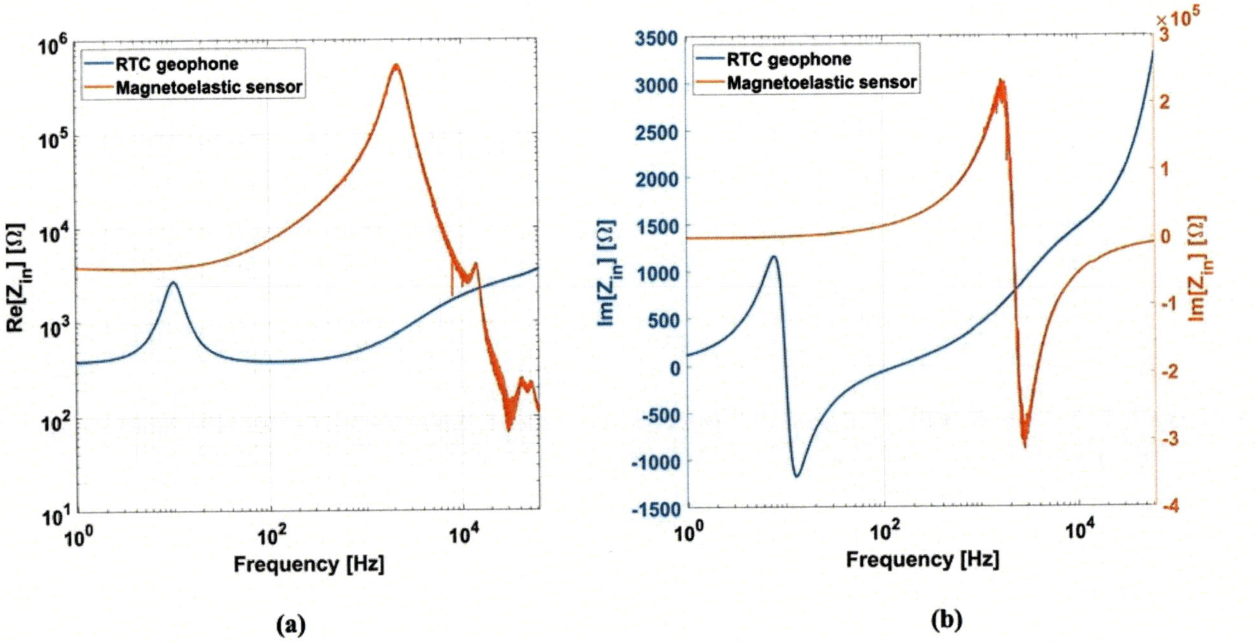

Fig. 4.3 The measured (**a**) real component and (**b**) imaginary component of the electrical impedance of the RTC 10 Hz geophone and the prototype magnetoelastic sensor

$$\text{MDS} = \frac{\text{Noise}\left[\frac{V}{\sqrt{Hz}}\right]}{\text{Sensitivity}\left[\frac{V}{m/s}\right]} \tag{4.2}$$

with units of $[m/s]/\sqrt{Hz}$. The predicted MDS is compared in Fig. 4.4. The geophone's simulated MDS is truncated at 250 Hz as this bounds the usable frequency range of the geophone due to spurious resonances. The magnetoelastic sensor is anticipated to have a broader bandwidth due to the stiffer suspension and excels in SNR performance at higher frequencies (i.e., above 150 Hz) owing to the +12 dB/octave increase in sensitivity with respect to the moving coil geophone. We emphasize that these are only predictions based off of measured impedance and measured sensitivity. Rigorous assessment of MDS is the subject of future work.

4.4 Conclusion

Here we presented basic characterization of a magnetoelastic inertial sensor prototype. The sensor has a native response to jerk, the time derivative of acceleration. As future work, we wish to explore the design space of this sensor concept using simulation techniques as we attempt to identify use-cases where these sensors may outperform other sensor types. In addition to the native jerk response, the sensors have a higher fundamental resonance frequency and, therefore, stiffer suspension compared to conventional geophones. This suggests that the technique may be well suited for broadband vibration sensors. Further, like the geophone but unlike piezoelectric inertial sensors, magnetoelastic sensors can be designed to have low output impedance and, therefore, can drive their signal down lengths of cable without the need for electronic amplification.

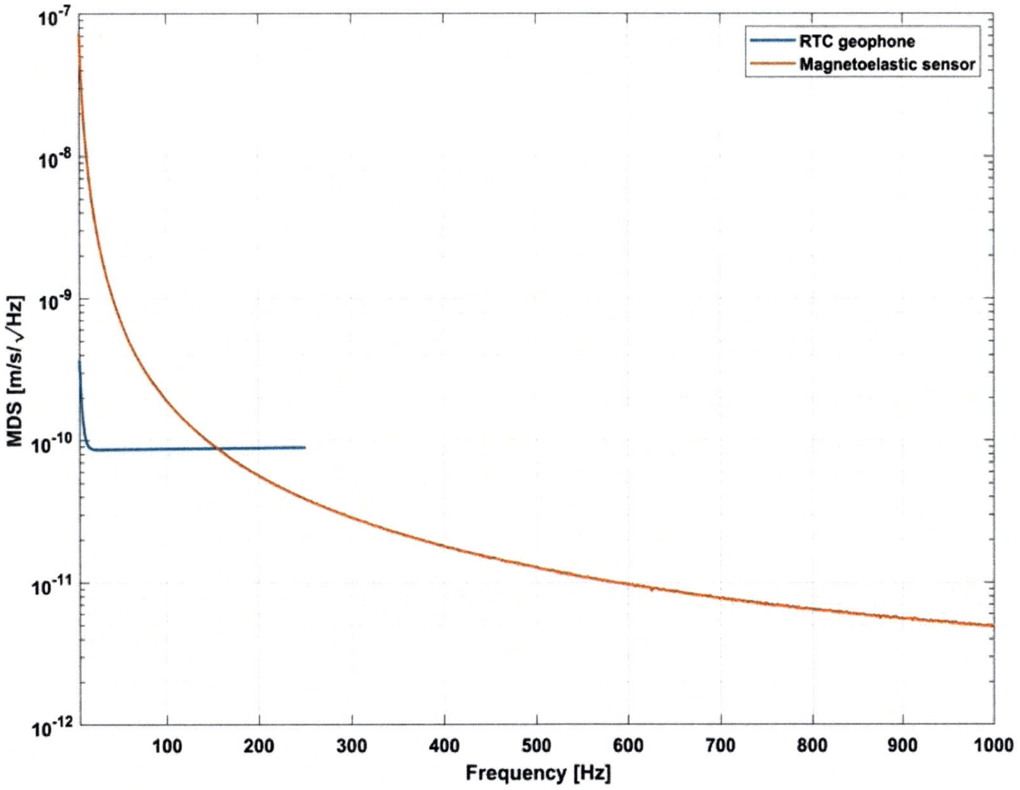

Fig. 4.4 Comparison of predicted minimum detectable velocity between the commercial RTC 10 Hz geophone and the prototype magnetoelastic sensor

References

1. Tumanski, S.: Handbook of Magnetic Measurements, pp. 131–135. CRC Press (2016)
2. Wei, Y., Yang, X., Chen, Y., Zhang, Z., Zheng, H.: Modeling of high-power Tonpilz Terfenol-D transducer using complex material parameters. Sensors. **22**(10), 3781 (2022)
3. Ren, L., Yu, K., Tan, Y.: Applications and advances of magnetoelastic sensors in biomedical engineering: a review. Materials. **12**(7), 1135 (2019)
4. Belyakov, A.S.: Magnetoelastic sensors and geophones for vector measurements in geoacoustics. Acoust. Phys. **51**(1), S43–S53 (2005)
5. https://www.rtclark.com/product/rtc-10hz-375-ohm-vertical-geophone-element-only/
6. Kim, D., Hewa-Kasakarage, N.N., Yoon, S.H., Kirk, K.D., Kuntzman, M., Hall, N.A.: Electrical admittance spectroscopy for piezoelectric MEMS. J. Microelectromech. Syst. **22**(2), 295–302 (2012)

Chapter 5
Implementation of Shaft-Mounted Accelerometer in the Local Fault Diagnosis of Geared Systems

Mohsen Azimi, Eniko T. Enikov, and Wyatt Pena

Abstract The goal of this chapter is to implement a wireless shaft-mounted accelerometer to provide a robust signal processing tool able to detect, localize, and assess mechanical irregularities in rotary machineries. For the purpose of experimental studies, a gear testing apparatus is designed and built to acquire vibration signal via both a traditional bearing-mounted accelerometer and the shaft-mounted accelerometer. First, a comparative analysis is outlined to compare the capability of the shaft-mounted and bearing-mounted accelerometers to the feature extraction of the damage. Then the quality of the vibration signal measured by the shaft-mounted accelerometer in different location and directions of the shaft is studied. Afterward, the sensitivity of the shaft-mounted accelerometer to the severity and location of the damage is evaluated by considering different size of one broken tooth and choosing different locations for the fault with respect to the transducer. Finally, the shaft-mounted accelerometer is used to obtain the same experimental results, while the speed is changing, which suggests the proposed approach is also effective towards condition monitoring of the gears under fluctuating load-speed condition.

Keywords Shaft-mounted accelerometer · Local fault · Equal angle resampling · Angle domain averaging · Variable operational condition

5.1 Introduction

Gears are the most important part of every power transmission system, designed to transmit motion from one shaft to another. A damaged gear tooth is a very common phenomena that mostly happens due to the fatigue, wear, extra load, strong impact, bad lubrication, corrosive environment, high temperature, backlash, eccentricity, etc., which results in excessive vibration, noise, waste of energy, and failure. During the last three decades, lots of effort have been directed towards condition monitoring and fault diagnostics of geared systems at early stages to prevent premature sudden breakdown and downtime of machineries or catastrophic malfunction. In most geared platforms, local faults appear as impacts, which changes the whole dynamic state of the machinery. To this end, vibration signals have been subject to intensive investigation and research since it contains rich information about meshing gears, shafts, bearings, and other moving mechanical components. Besides, it can be measured via accelerometers during the regular operation of the system and without stopping or interrupting the machinery.

However, vibrational signals are heavily corrupted by measurement noise and due to the low signal-to-noise ratio (SNR) nature of the signal, the transitory information of the defect is almost masked by the noise, making it difficult to extract key features and faults. Time Synchronous Averaging (TSA) is the most common denoising technique that is employed to improve SNR by averaging the periodic sections of the signal in the time domain [1–3]. TSA can remove the background

M. Azimi (✉)
Department of Mechanical Engineering, Mississippi State University, Starkville, MS, USA
e-mail: azimi@me.msstate.edu

E. T. Enikov
Department of Aerospace and Mechanical Engineering, University of Arizona, Tucson, AZ, USA
e-mail: enikov@arizona.edu

W. Pena
Ridgetop Group Inc., Tucson, AZ, USA
e-mail: wpena@ridgetopgroup.com

C. Walber et al. (eds.), *Sensors & Instrumentation and Aircraft/Aerospace Testing Techniques, Volume 8*, Conference Proceedings of the Society for Experimental Mechanics Series, https://doi.org/10.1007/978-3-031-34938-6_5

noise and any periodic events not exactly synchronous with the period of the gear of interest, which makes it a strong tool for fault detection and localization of multistage gearboxes [4]. But the underlying assumption for this method is constant operational speed, which limits TSA only to stationary signals. This is while due to the friction, uneven mass distribution, gear teeth transmission error, and uncertainties associated to the power and load, most of the geared systems operate under fluctuating load and speed conditions [5], which makes the obtained vibrational signal nonstationary. In such systems, fixed time sampling is like choosing a variable sampling rate frequency analysis that results in leakage error and spectral smearing [6–8]. To resolve this problem, Order Tracking Methods are developed that try to analyze the signal in order (angular) domain rather than time domain. The angular domain sampled signal can be provided in two different ways; first, sampling the vibration signal at constant angular increments at a rate proportional to the shaft speed known as Instrumented Order Tracking. Second is sampling the vibration signal at constant time increments and then resampling the time domain signal to provide the angular domain data known as Computed Order Tracking [9, 10]. Independent Angular Resampling (IAR) is a very common Computed Order Tracking approach that involves analysis of vibration signal under either speed-up [7] or run-down conditions. In this method, the angular acceleration is assumed to remain constant during each independent revolution, and the angular position is estimated through fitting a quadratic polynomial based on the angular position at three different time instants [6, 10–12]. Nevertheless, IAR cannot be used for systems with large speed variations [8]. Consequently, Speed Based Resampling Scheme (SBRS) is another Computed Order Tracking approach, mostly used for systems with variable acceleration. In [13], a one-pulse tachometer is used to estimate the average rotational speed within a rotation and the amplitude of the vibration is determined via speed based resampling interpolation. [11] tried to improve the accuracy of this approach by implementing a three-pulse tachometer to account even for speed fluctuations within a single rotation. [1] uses an n-pulse tachometer for angular domain resampling interpolation with constant angle increments to ensure a constant integer number of samples per revolution under any speed fluctuations. In all aforementioned works, after obtaining the constant angular increments signal by resampling the recorded constant time increment vibration signa, the Angular Domain Averaging (ADA) technique is employed to remove the background noise and the periodic events that are not synchronous with the gear of interest.

There are many other works that used the frequency spectrum of the vibration signal to study the health condition of the gears [14, 15], which is mainly done by classifying faults into localized and distributed irregularities [16]. Usually, the local irregularities start with scuffing and pitting on the face of operating gears, causing noise and excessive vibrations. As the time evolves, the pitting faults extend to the spalling faults and later lead to a cracked or broken gear tooth fault, which can cause catastrophic failure. Generally, localized faults cause change in the gear meshing stiffness and, consequently, periodic short-time-duration impulses, which have relatively low amplitude with wide-band frequency. The local fault has a serious effect on the shaft's frequency, but little influence on the Gear Meshing Frequency (GMF) [16]. On the other hand, gear faults like wear, tear, pitting, peeling, and misalignment result in change of the gear transmission error and are considered as distributed faults which have high amplitude short-band spectrum. Distributed faults increase the number and amplitude of the GMF sidebands [9], which are considered as a useful indicator to locate the faulty tooth in platforms with multiple shafts. Even though the frequency spectrum of vibration signals is sensitive to possible irregularities, it is not the best tool for extracting detailed information on the transitory events, which is the main characteristic of faults in gears [17, 18]. Furthermore, this method is unable to detect gear failures at early stages [19, 20].

In recent years, the time-frequency analysis techniques, including Short Time Fourier Transform (STFT) [21] and Continuous Wavelet Transform (CWT) [22], have become popular as multiresolution tools for extracting the short-duration components of vibration signals. It has been shown that even though STFT is able to distinguish the presence of the fault in a faulty gearbox compared to a healthy gearbox, still it is not suitable for analyzing gearbox vibration signals that contain multiscale components. This is mainly because STFT has a constant time-frequency resolution in the entire time-frequency plot [3, 9]. The complex Morlet Mother Wavelet (MMW) is the most popular mother wavelet used in CWT for the application of local fault diagnostics in gears. This is mainly because of the great similarity between the MMW and the periodic short-time-duration impulses in the vibration signal, which are caused by the local faults [1, 12]. It is also worth noting that the shape and size of the MMW can easily be adjusted according to the features of the expecting transient impulsive components [23, 24]. The best result is obtained when the MMW matches the time-frequency structure of the transient impulses [24]; therefore, several researchers tried to optimize the MMW parameters to enhance the accuracy of fault detection [22, 25, 26]. Accordingly, the Polar Wavelet Mapping (PWM) method is introduced to improve the visual representation of CWT by the display of the amplitude versus the rotational angle in polar coordinates [27]. Some researchers used a combination of the above-mentioned denoising and feature extraction methods. In [23], a new method is proposed based on combining developed time averaging and adaptive Morlet wavelet.

The primary objective of this chapter is to use a shaft-mounted accelerometer to collect more enriched vibration signal which results in a more sensitive and reliable health monitoring process [28, 29]. To this end, a comparative analysis is performed to demonstrate the effectiveness of the shaft-mounted accelerometer compared to the traditional bearing pedestaled accelerometer. To present a comprehensive analysis, the local fault diagnostics are performed in order-frequency domains for different level and location of induced local fault. Additionally, unlike the traditional bearing-mounted accelerometer, the shaft-mounted accelerometer is able to collect data in tangent, transversal, and longitudinal directions of the shaft. Consequently, collecting the vibration signal in different directions and drawing a comparative analysis between them is another topic covered in this work. At the end, it is demonstrated that by implementation of 16 magnets for the tachometer, the local fault can be detected even under fluctuating load-speed conditions.

The rest of this chapter is organized as follows: First, the theoretical technique, which is employed in the present work to detect and assess the simulated local faults, is explained. Then the experimental test rig and the wireless data acquisition system is demonstrated. Afterward, the experimental results are represented. At the end, the conclusion and remarks are presented.

5.2 Implemented Fault Diagnostics Approach

The algorithm explained in this section is used for the local fault identification of the gear set demonstrated in the following section. In all of the case studies provided in this work, SBRS is first used to convert the measured time domain vibration signal into angular domain signal where a fixed number of digital sampling data are carried out at equal angular displacements of the gear over one complete rotation. Next, FFT is used to find the frequency spectrum of the resulting signal for the investigation of the harmonic components. Then, the residual signal is obtained by removing the shaft frequency from the frequency spectrum, where the angular domain residual signal is averaged over 20 revolutions to obtain the averaged residual signal. Afterward, MMW with varying center frequency and constant dilatation parameter is used for CWT to pinpoint the angular location and the frequency of the periodic impulsive components of the averaged residual signal. Finally, the result is plotted in the polar form to provide better visual demonstration. Figure 5.1 demonstrates an illustrative description of the above-mentioned algorithm.

5.3 Speed-Based Resampling Scheme (SBRS)

In this work, a 16 pulse per revolution tachometer is used to provide a relatively accurate resolution of the angular position (fixed order) such that the time signal between the two tachometer pulses is considered as 22.5° revolution of the output shaft and the sampling rate vary over time with change of the shaft's rotational speed. Consequently, by assuming constant angular velocity between the two tachometer pulses, linear interpolation is used to map the time domain signal into angular domain

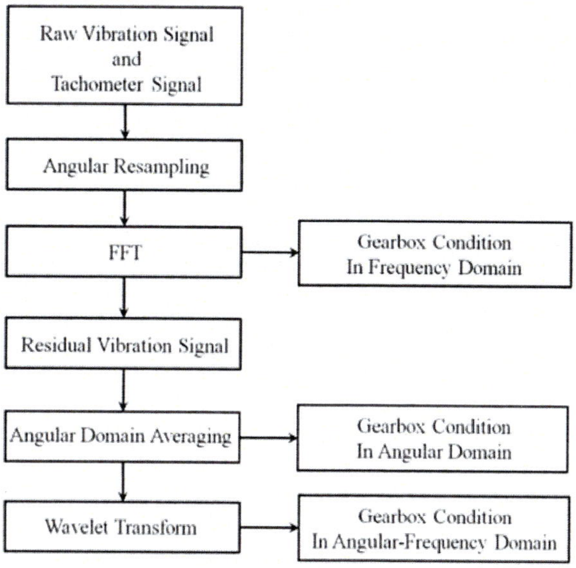

Fig. 5.1 Flowchart of the implemented local gear fault diagnostics approach

signal. Additionally, the full circle is divided into the number of the gear tooth and a 0° reference position is specified to provide a precise localization of any fault on the gear.

5.4 Residual Signal

In this work, the residual signal is obtained by elimination of the shaft frequency from the frequency spectrum of SBRS signal. The residual method aims to extract the weak harmonic short duration components [30] and reveal the presence of local fault clearer than the overall vibration signal [27]. Notice that the range of the frequency content to be removed must be adjusted such that the frequency spectrum related to the periodic short-term duration impulses are not removed from the signal.

5.5 Angular Doman Averaging (ADA)

The frequency domain analysis is a potential technique to identify the basic characteristics of any vibration signal, like shafts frequency, meshing frequency, their harmonics and subharmonic. At this stage, investigating the sidebands of the shaft and meshing frequencies can provide a good sense on the existence of the local and distributed faults. In this work, the vibration signal is denoised via ADA such that the angular synchronous signal is averaged over several number of revolutions [6, 11]. This lets us utilize all the benefits of TSA for the noise suppression and to avoid spectral leakage and smearing in the frequency spectrum.

$$\overline{x}(\theta) = \frac{1}{N}\sum_{n=0}^{N-1} x(\theta - nT) \tag{5.1}$$

where $x(t)$ is the raw signal, T is rotational period and N is the total number of averages.

5.6 Continuous Wavelet Transform (CWT)

CWT is a linear transformation very similar to the FT, except that instead of sine and cosine functions, Mother Wavelet Function (MWF) is used as the elementary basis function. MWF is a series of windowing functions with variable frequencies.

$$W = \int_{-\infty}^{+\infty} x(t)\Psi^*(t)dt \tag{5.2}$$

where $x(t)$ is the signal and $\Psi^*(t)$ is the conjugate of the kernel function $\Psi(t)$. There are different types of real or complex MWF. The best fault feature extraction is obtained when the utilized MWF has the greatest similarity to the transient components of the signal. In this work, the complex Morlet Mother Wavelet (MMW) is employed owing to its similarity to the periodic transient components in acquired vibration signals, which is the characteristic of local faults in geared systems [27]. This is mainly because the local fault reduces the meshing stiffness of the damaged tooth, which results in change of the periodic excitation; besides, the stiffness between the neighboring teeth will be influenced by the fault, and due to the resulting time delay in the excitation frequency, the fault appears as a harmonic short duration component in the vibration signal [31].

$$\Psi_{a,b}(\mathrm{t}) = \sqrt{\mathrm{b}}\, e^{i\omega_0(t-a)} e^{-\frac{b}{2}(\mathrm{t}-a)^2} \tag{5.3}$$

where a is the translation factor, b is the dilation factor, ω_0 is the center frequency and $\sqrt{b}$ is the energy normalization factor which scales the energy of the wavelet for different dilation factors. MMW is a complex exponential function with a helical three-dimensional form and a harmonic two-dimensional form on real and imaginary planes [32], as demonstrated in Fig. 5.2.

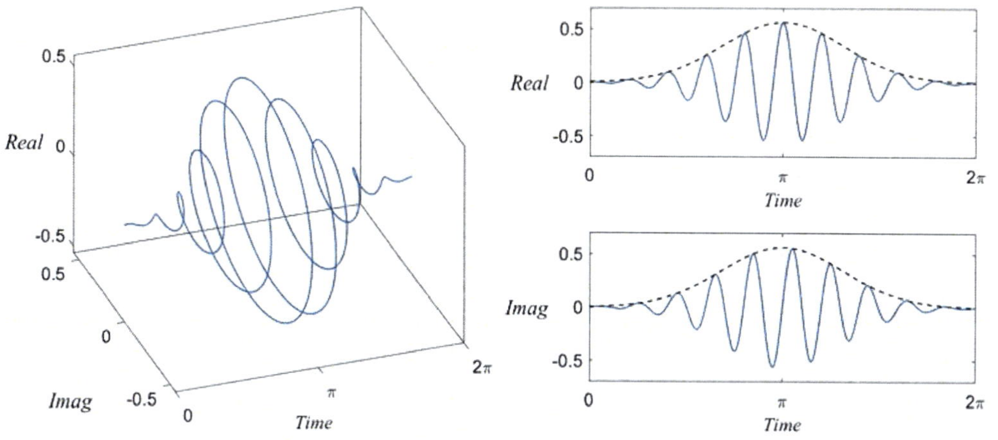

Fig. 5.2 MMW for $a = \pi, b = 1, w_0 = 10$

Choosing suitable parameters for MMW is an empirical design problem and depends on the signal characteristics that result in a good balance of time and frequency resolution. The resulting CWT is obtained by inserting Eq. 5.3 in Eq. 5.2.

$$W_{a,b} = \sqrt{b}\int_{-\infty}^{+\infty} x(t)e^{-i\omega_0(t-a)}e^{-\frac{b}{2}(t-a)^2}dt \tag{5.4}$$

In fact, Eq. 5.4 can be considered as convolution between the signal and the conjugate of the dilated-translated MWF, which reflects the correlation between them:

$$W_{a,b} =< x(t), \Psi^*_{a,b}(t) > \tag{5.5}$$

Alternatively, this convolution integral can be calculated as a simple inner product in frequency domain, which has been found beneficial in reducing the computational cost and providing a better illustrative description for the change of the dilation factor and center frequency [32]:

$$W_{a,b} = \text{IFFT}\left\{\text{FFT}\left(x(t)\right) * \text{FFT}\left(\Psi^*_{a,b}(t)\right)\right\} \tag{5.6}$$

where IFFT denotes the Inverse Fast Fourier Transform. Figures 5.3 and 5.4 illustrate the effect of changing the dilation factor and center frequency on the spectrum of MMW in time and frequency domains. Figure. 5.3 shows the FFT of the MMW for different center frequencies and constant dilation factor, while Fig. 5.4 shows the FFT of MMW for constant center frequency and different dilation factors. A simple comparison between Figs. 5.3 and 5.4 shows that increasing the value of the center frequency shifts the location of the frequency spectrum toward higher frequencies, while increasing the value of the dilation factor decreases the width of the frequency spectrum. In other words, the value of the center frequency determines the location of the frequency spectrum, and the dilation factor determines the resolution of the frequency spectrum such that increasing the value of the dilation factor increases the resolution in frequency domain and decreases the resolution in the time domain and vice versa.

5.7 Experimental Setup

Figure 5.5 shows a schematic of the experimental setup of a single stage spur gear arrangement utilized in this work, in which a 15-tooth driver pinion is in mesh with a 32-tooth driven gear. Both gears are made of steel with pressure angle of 15.4 degrees, and gears are in mesh without any lubrication. This gear set has two shafts, both made of steel and supported by two pillow ball bearings such that half an inch diameter input shaft is driven by a 0.5 HP, 0–1500 rpm variable speed DC motor through pulley and V-shaped belt and the one-inch diameter output shaft is connected to a Yamaha YZ250F motorcycle front disk break with 4–220 lb.in capacity break through pulley and V-shaped belt, which enable us to run the system under different load-speed conditions.

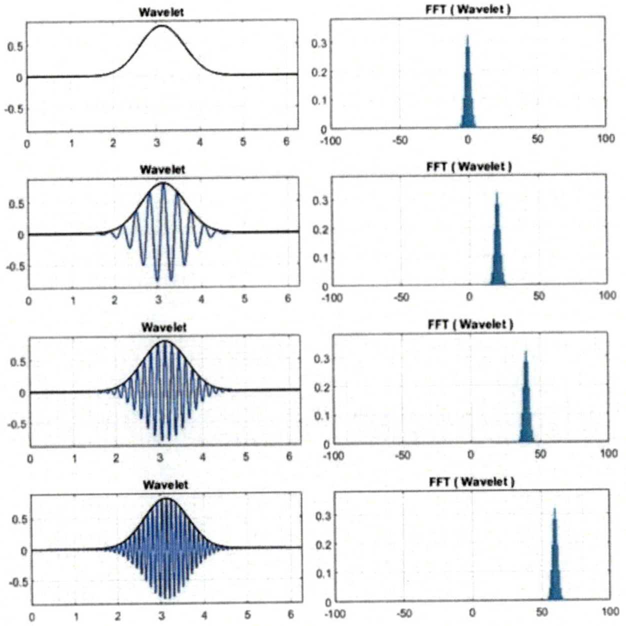

Fig. 5.3 b = 0.5, ω0 = 0, 20, 40, 60 rad/sec

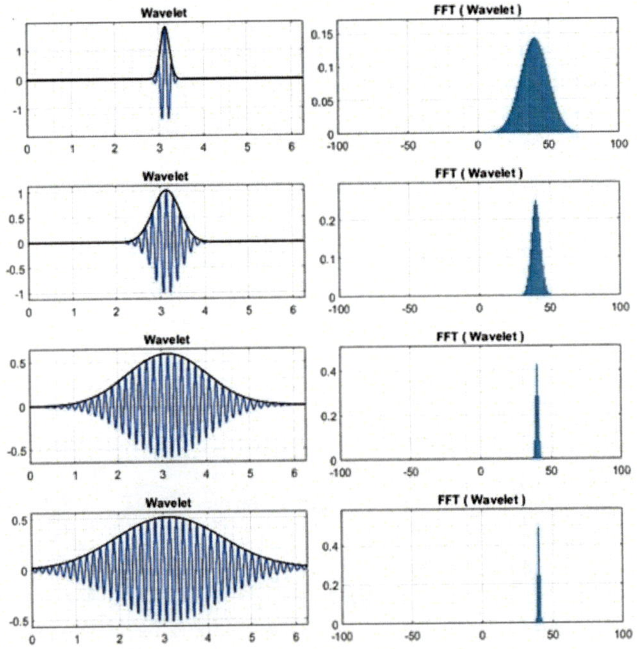

Fig. 5.4 b = 0.1, 0.3, 0.9, 1.2, $\omega_0 = 40$ rad/sec

Two identical three axial analog MEMS accelerometers, EVAL-ADXL325EB, with sensing range of ±6 g and sensitivity of 174 mV/g, are used to measure the vibration signal, one mounted directly on the shaft, Fig. 5.6a, and one mounted on the outer surface of the output shaft's bearing, Fig. 5.6b. The vibration signal is acquired with the sampling frequency of 40 kHz, in a 32-bits one channel ADC data acquisition system. The shaft-mounted accelerometer enables us to collect the vibration signals from different locations along the longitudinal, transversal, and tangent directions of the shaft. This is while because of the similarity between the acceleration signals obtained be the bearing-mounted accelerometer in both vertical and horizontal directions, there is no clear advantage when the vibration signatures is acquired in two mutually perpendicular directions [11].

As it is demonstrated in Fig. 5.7, frames for 16 magnets are mounted on the bearing and the shaft for the shaft mounted and bearing mounted tachometers, respectively. For both cases, the 0° reference position is specified by removing one of the

Fig. 5.5 Schematic of the experimental setup

a. Mounted on the shaft **b. Mounted on the bearing**

Fig. 5.6 MEMS accelerometers

magnets. The data acquisition system for this experiment was sponsored by Ridgetop Group using a customized Sentinel Motion Development Kit (SMDK). Ridgetop's Sentinel Motion technology originated through a NASA SBIR program and is an Internet of Things (IoT)–based sensor system that monitors mission critical equipment observing any combination of temperature and linear, rotatory, or vibrational force. Sentinel Motion comprises a wireless network of RotoSense smart sensors, the Sentinel Gateway communications device, and the Sentinel MotionView software package for data acquisition, analysis, and sensor-gateway management.

Each RotoSense unit is able to collect data in three channels of ADC at the same time. In this work, only two of these channels are used; one for the analog accelerometer and one for the optical tachometer. The collected data are streamed to the Sentinel Gateway through the low powered IEEE 802.15.4 wireless radio frequency communication protocol. The transmitted data are then accessible via a network drive that is provided by the Sentinel Gateway access point. This allows exporting the data to computing platforms such as Microsoft Excel and MATLAB (Fig. 5.8).

Fig. 5.7 Frames for magnets of the tachometer

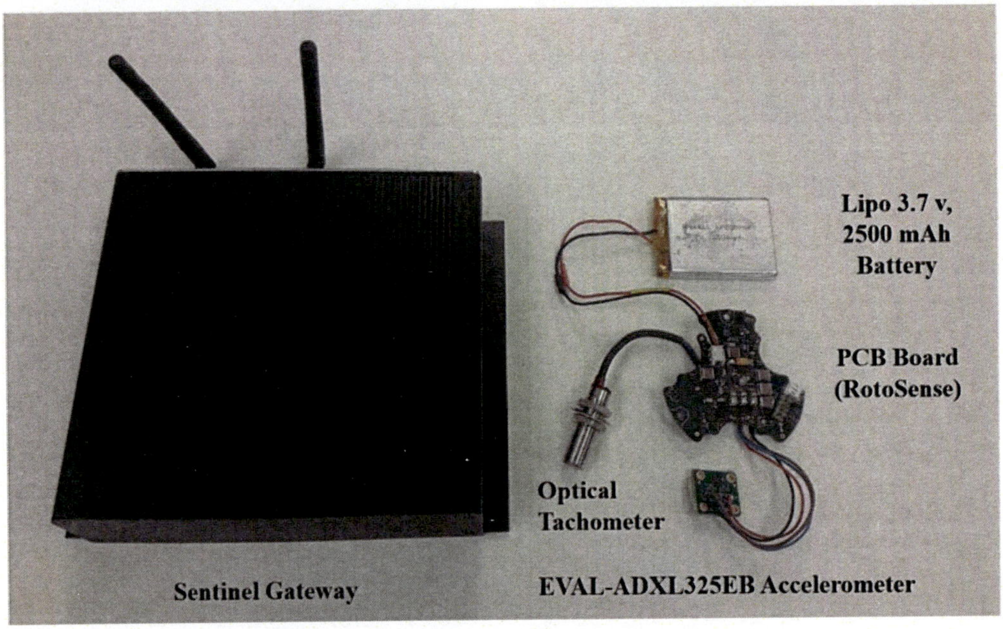

Fig. 5.8 PCB board and Sentinel Gateway

5.8 Results

For each experiment presented in this work, 2 s length of the vibrational signal is measured and analyzed, while the system in operating around 1000 rpm, that is, 16.67 Hz, which results in 20 complete revolution of the output shaft. Consequently, MATLAB programing is used for the signal processing and extraction of diagnostic features according to the flowchart provided in Fig. 5.1. Figure 5.9 shows the views of the three gears with different size of seeded faults, which are introduced via machining and used in this experiment. In the subsection "A," the result of the signal processing is provided for each stage, as described by Fig. 5.1, but for the rest of the subsections, only the final result of the signal processing obtained by CWT is provided, compared, and studied (Figs. 5.10 and 5.11).

5.9 Shaft-Mounted Accelerometer Versus Bearing-Mounted Accelerometer

In this section, the vibration signal measured by the shaft-mounted and bearing-mounted accelerometers are analyzed, once for the healthy pinion and faulty gear condition and once for the healthy pinion-gear condition. The results are compared to demonstrate the effectiveness of each case in the local fault diagnostics of gears. In Figs. 5.12, 5.13, 5.14 and 5.15, section

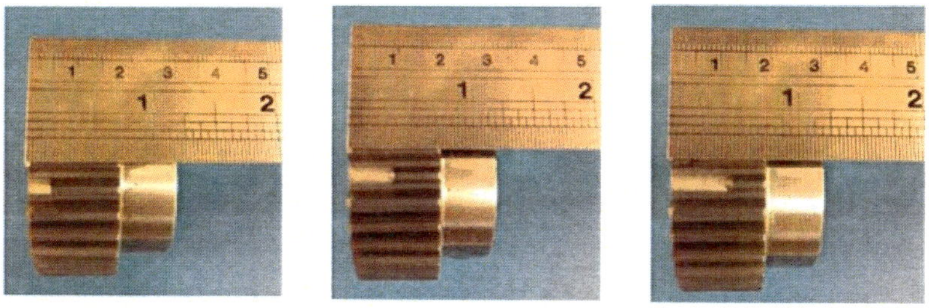

25% broken tooth 50% broken tooth 65% broken tooth

Fig. 5.9 Gear with different size of broken tooth

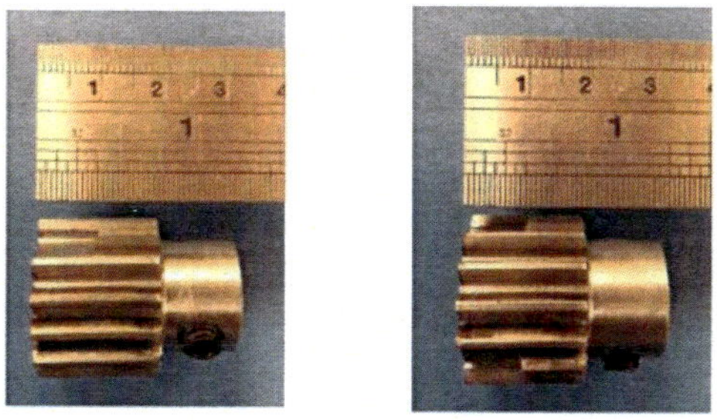

One 50% broken tooth Two 50% broken teeth

Fig. 5.10 Pinion with different size of broken tooth

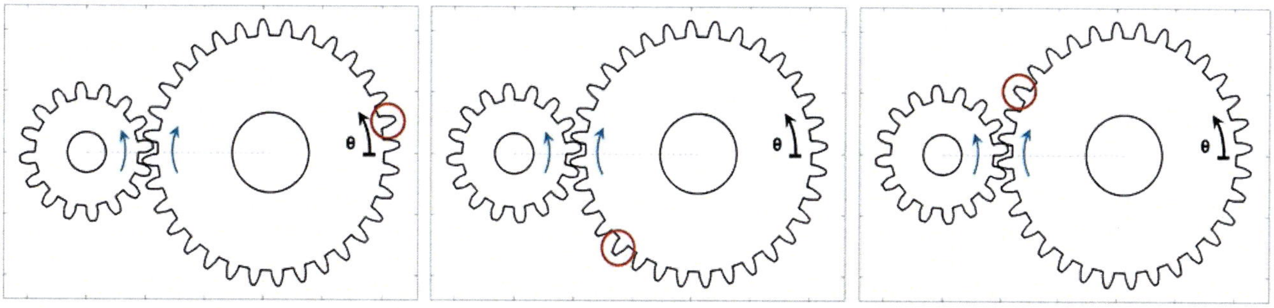

a. One gear teeth damaged around θ≈10 (deg) **b.** One Gear teeth damaged around θ≈150 (deg) **c.** One gear teeth damaged around θ≈230 (deg)

Fig. 5.11 Different location for the fault

a shows the measured raw vibration signals over 20 rotations of the output shaft and section b demonstrate the FFT of this signal. Section c shows the FFT where the low frequency components, below 45 Hz, of the signal is removed. Section d shows the residual signal where IFFF is used to transform the filtered signal into time domain. Section e, demonstrate the averaged residual signal in the angular domain. And at the end the result of the CWT is provided in the polar coordinate to demonstrate the angular location of local fault. Comparing section b in Figs. 5.12, 5.13, 5.14 and 5.15 shows that in the faulty condition there exists an increase in the magnitude of the shaft's frequency which is an indication of defective condition of gear tooth. As it can be seen in section e of Figs. 5.7 and 5.8, the short duration harmonic component caused by the local fault is masked by the high frequency noise in the signal. This is where the CWT is used to extract this component from the averaged residual signal. Finally, comparing the results obtained in frequency, time, and time-frequency domains show that the result obtained by the shaft-mounted accelerometer is as effective and powerful as bearing-mounted accelerometer for capturing periodic short-term components caused by local faults.

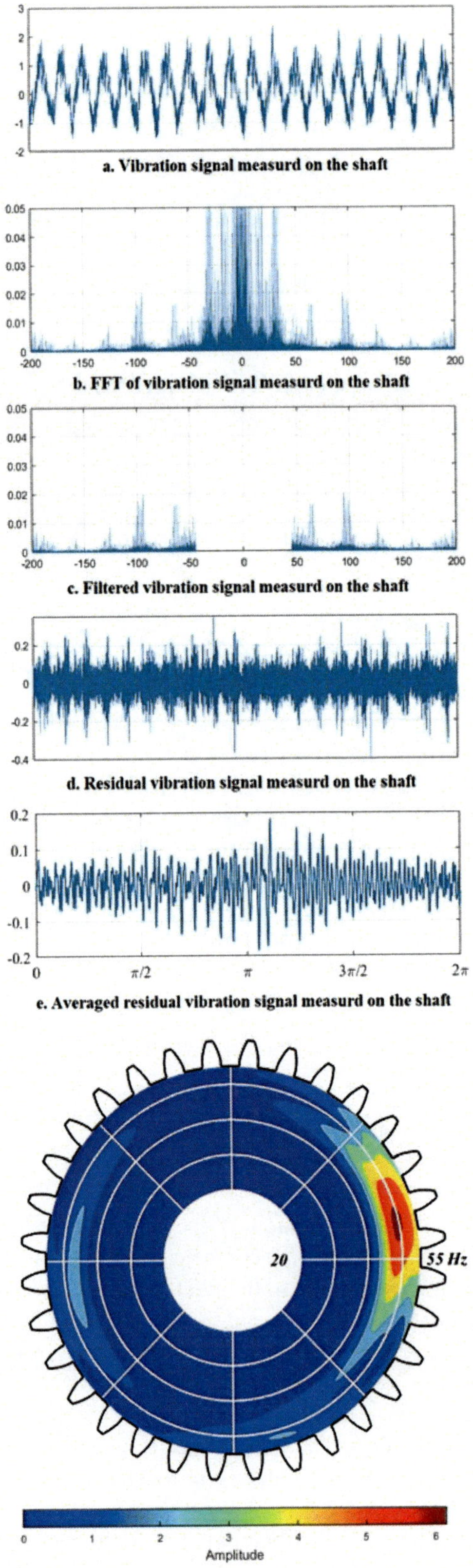

Fig. 5.12 Faulty case for the shaft-mounted accelerometer

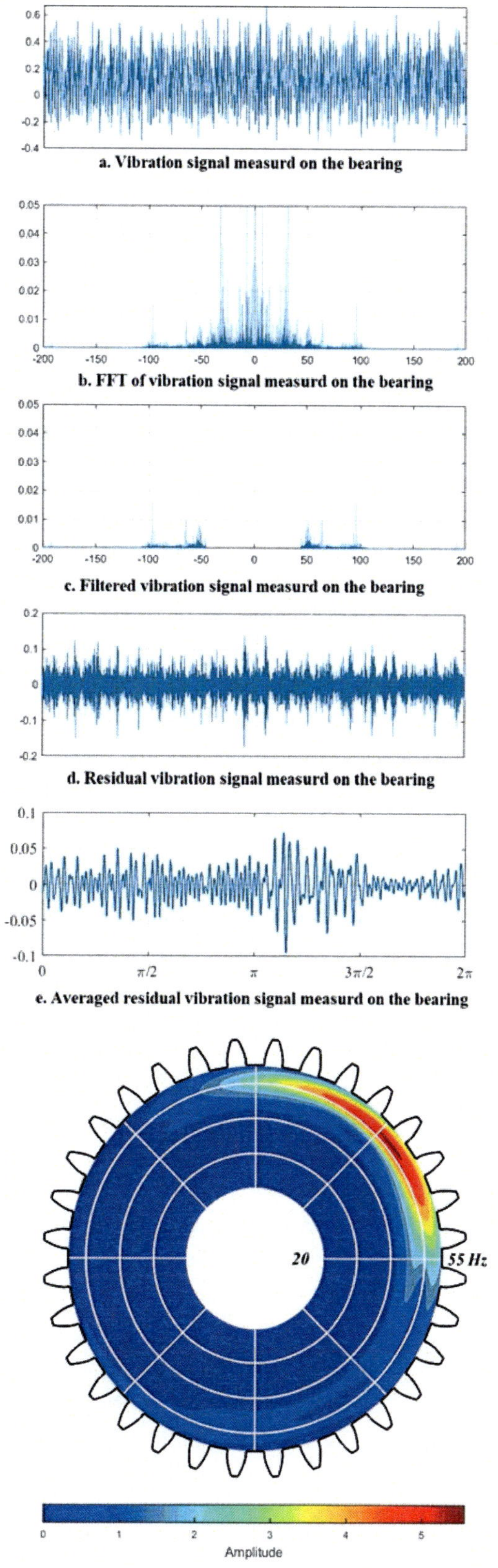

Fig. 5.13 Faulty case for the bearing-mounted accelerometer

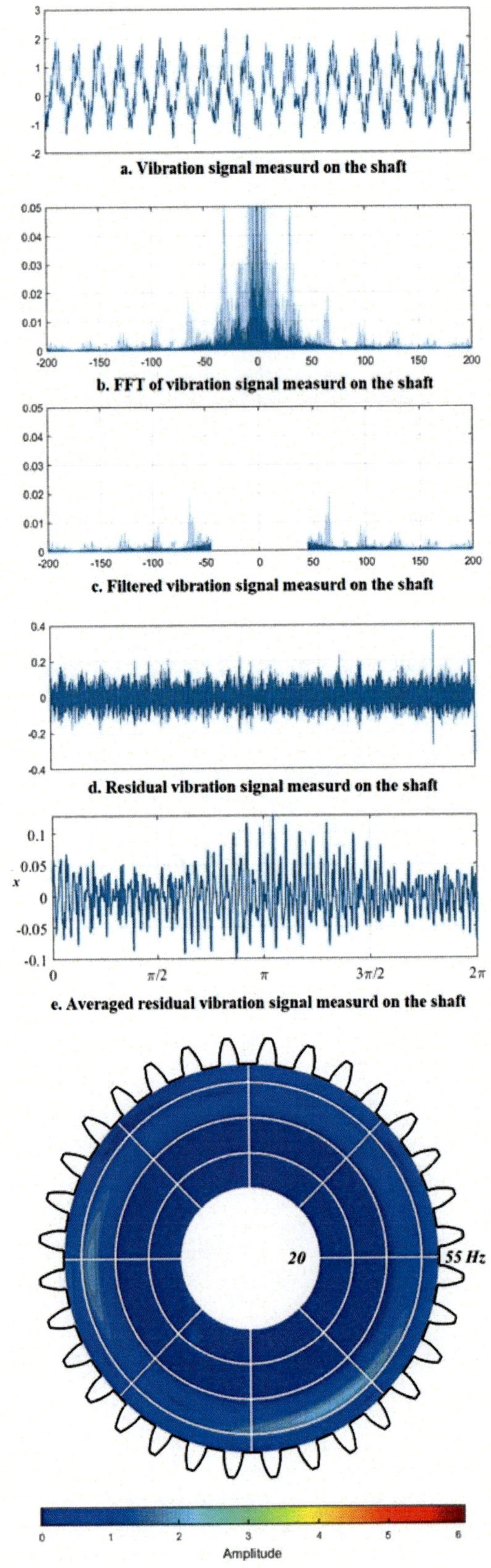

Fig. 5.14 Faulty case for the shaft-mounted accelerometer

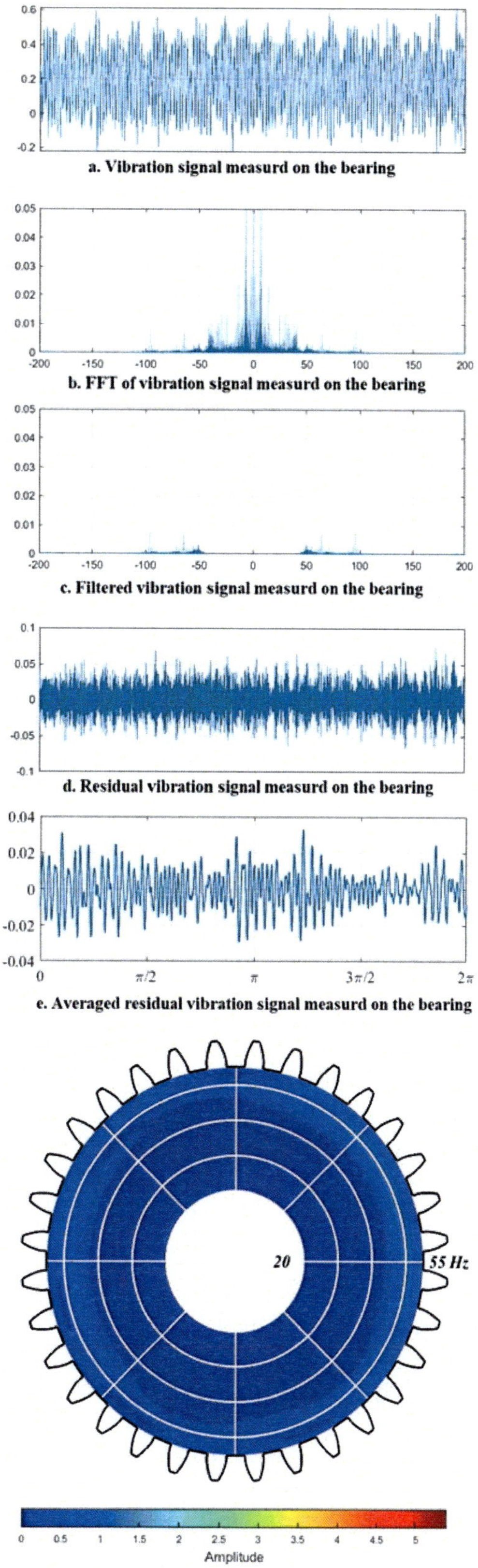

Fig. 5.15 Faulty case for the bearing-mounted accelerometer

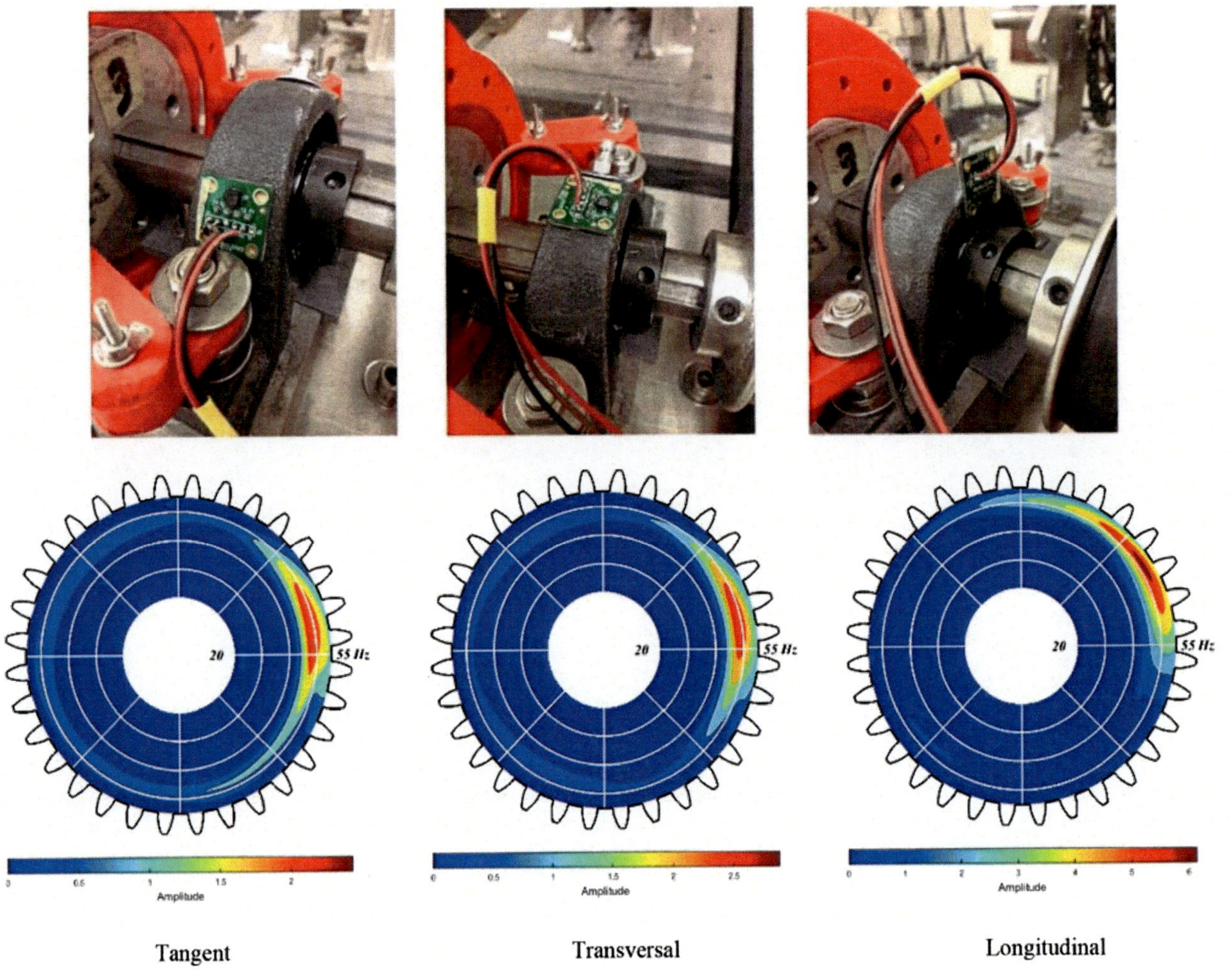

Fig. 5.16 Vibration signal measured in different directions on the pillow ball bearing

5.10 Measurement from Different Locations and Directions on the Pillow Ball Bearing

In this section, the vibration signal is measured by mounting the accelerometer on different location of the output shaft's bearing. This experiment show that the vibration signal measured in the longitudinal direction of the pillow ball bearing provide the best result where the largest amplitude is obtained for the short-time harmonic term (Fig. 5.16).

5.11 Measurement from Different Locations and Directions on the Shaft

Implementation of a shaft-mounted accelerometer enables us to measure the vibration signal in different locations and directions of the shaft. Therefore, in this section a comparative study is performed to demonstrate the enrichment of the vibration signal measured in different location and directions on the shaft. The results of all the case studies are demonstrated in Figs. 5.17, 5.18 and 5.19. The goal is to determine at which coordinate and orientation the periodic short time component of the vibration signal can be captured better. Figure 5.17 shows that measuring the vibration signal in the transversal direction at different locations anywhere between the center and the perimeter of the shaft provide the same result. This is while. The transversal vibration signal measured on the surface of the shaft do not provide any indication of presence of the fault. Figure 5.18 shows that by measuring the vibration signal in the longitudinal direction and by moving the location of the accelerometer from the center to the surface of the shaft better result is obtained as the accelerometer get closer to the surface of the shaft. Figure 5.19 shows that measuring the vibration signal in tangent direction results in a better fault diagnostic

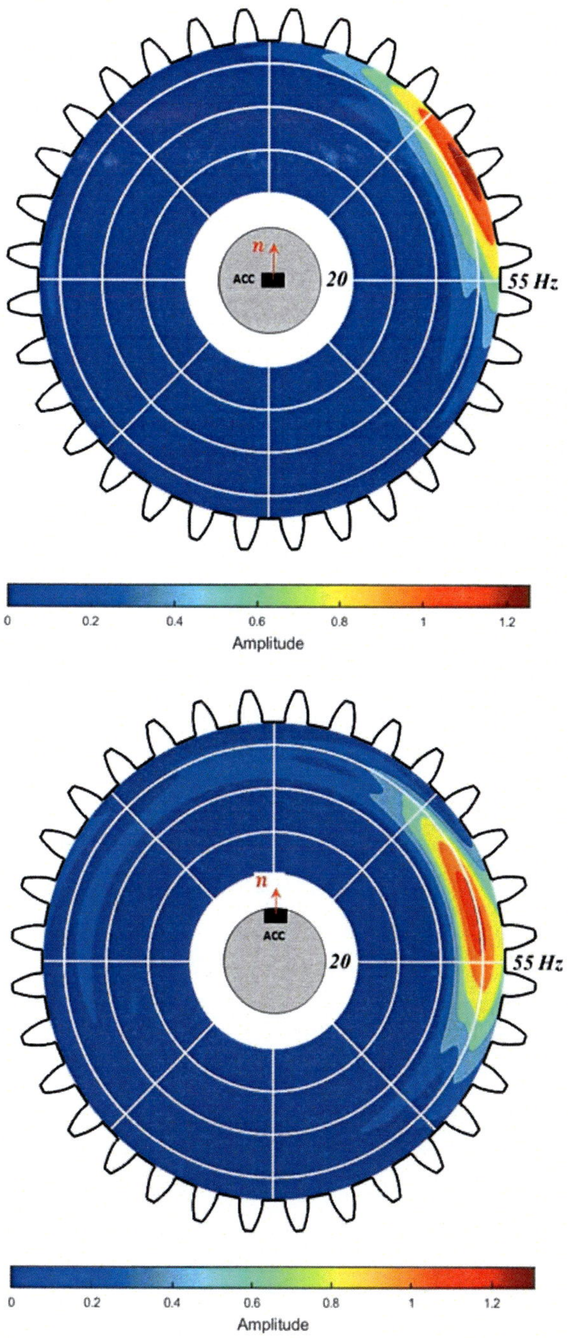

Fig. 5.17 Results from vibration signal measured in the transversal direction for different location of the accelerometer

when the accelerometer is mounted on the surface of the shaft. Finally, comparing the best results obtained in Figs. 5.17, 5.18 and 5.19 shows that the tangent direction is the best direction to measure the vibration signal.

5.12 Different Locations and Size for the Fault

In this section the sensitivity of the results obtained by CWT to the dimension of the broken tooth is investigated, where a healthy pinion and faulty gears with 25% 50% 65% broken tooth conditions are used, illustrated in Fig. 5.9. To this end, the change in the location of the faulty tooth is achieved by mounting the tachometer in three different angular location with

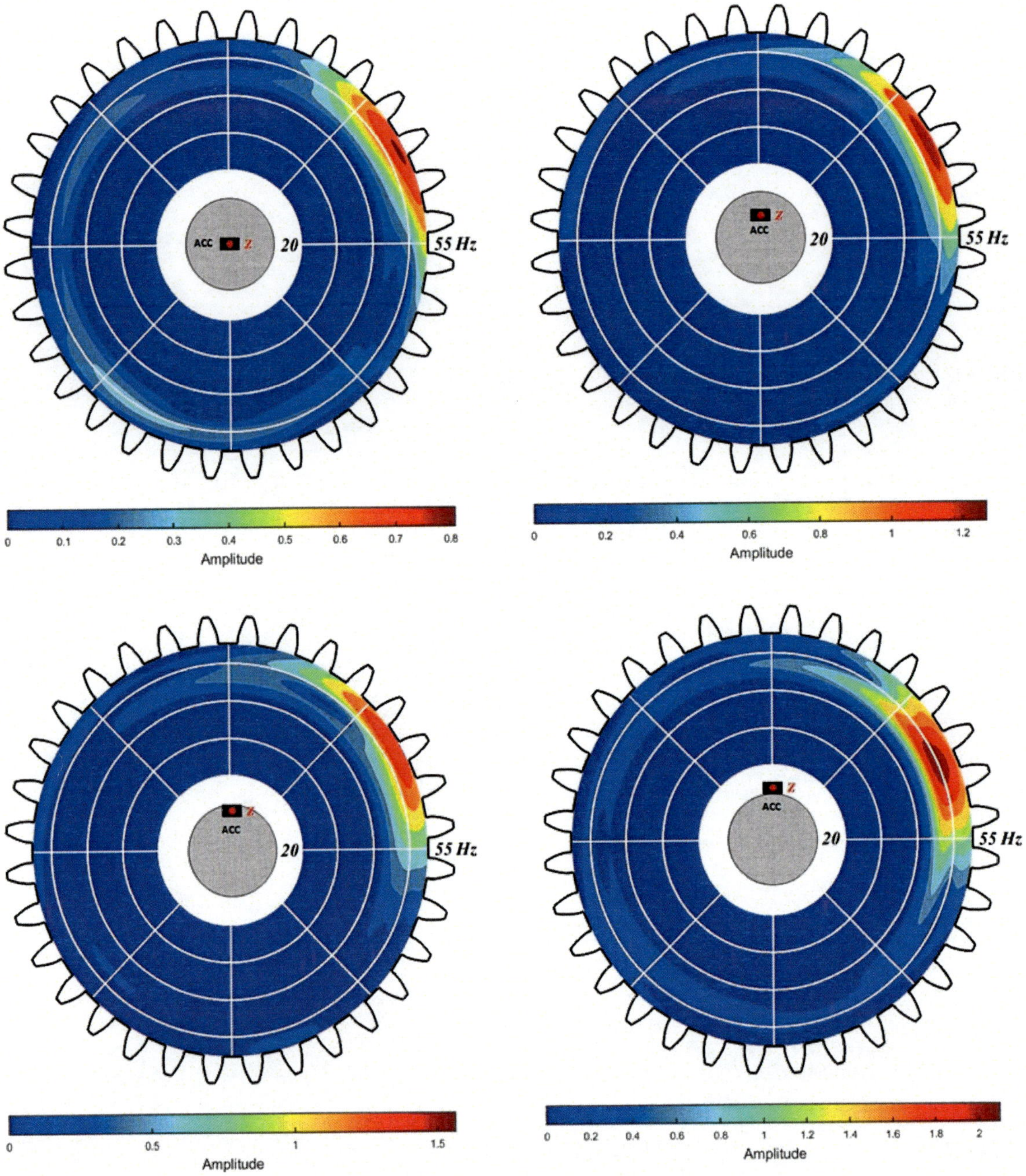

Fig. 5.18 Results from vibration signal measured in the longitudinal direction for different location of the accelerometer

respect to the faulty tooth, illustrated in Fig. 5.10, and the shaft-mounted accelerometer is used to detect, localize and asses the fault in both longitudinal and tangent directions. Figures 5.20, 5.21, and 5.22 show that the implemented method is able to determine the location of the fault. This is while the increase in the severity of the damage case increase in the amplitude of the short-term harmonic component, which is caused by the local fault of the gear.

5.13 Faulty Gear Verses Faulty Pinion

In this section, the fault diagnostic of faulty pinion, demonstrated in Fig. 5.10, and healthy gear condition is investigated. Figures 5.23 and 5.24 show the result related to pinion with one and two damaged teeth, respectively. For each case, ADA

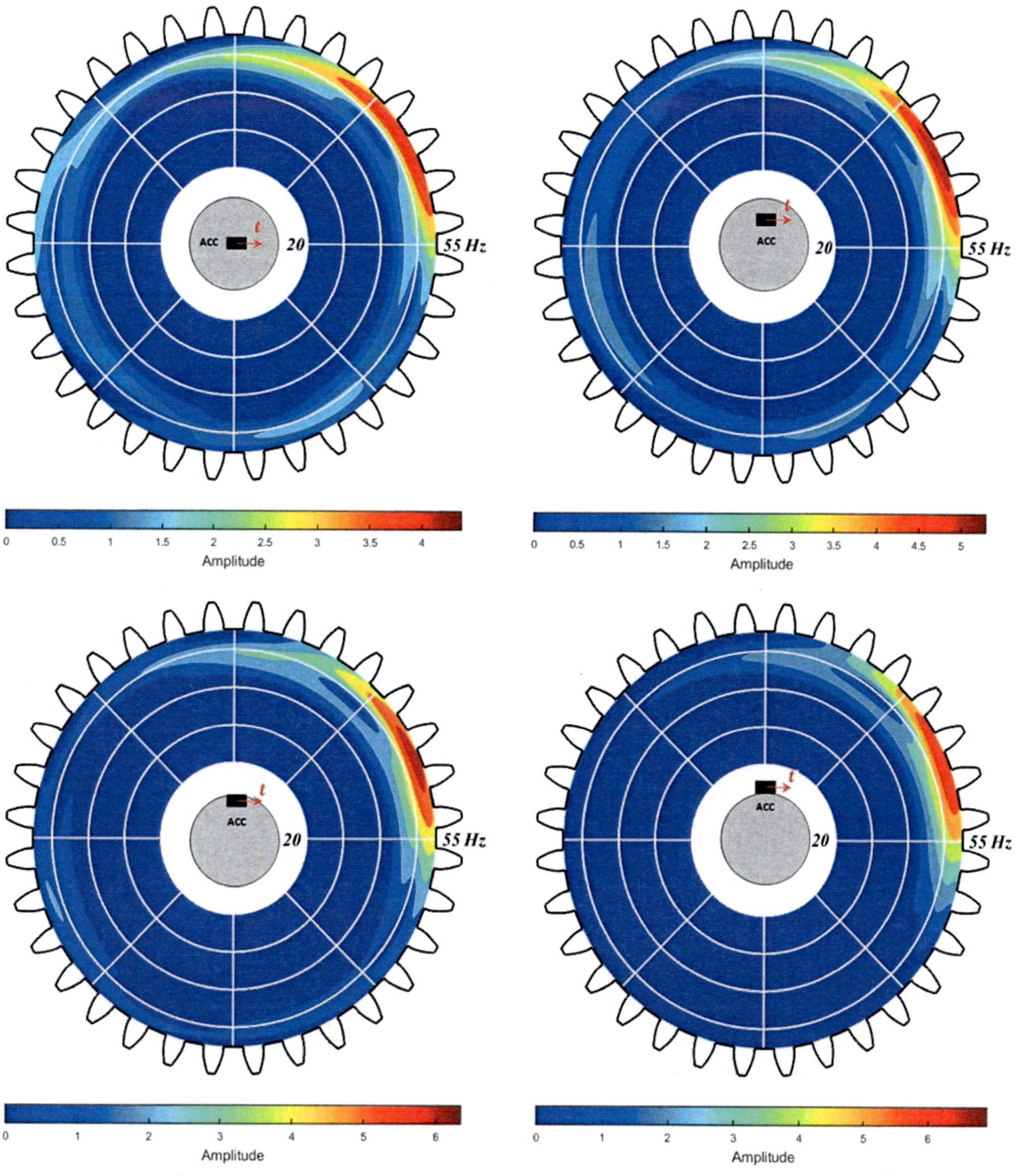

Fig. 5.19 Results from vibration signal measured in the tangent direction for different location of the accelerometer

is performed such that the speed of the pinion and gear are chosen as the base frequencies to identify and assess the severity of the seeded fault located on pinion and gear. Figures 5.23 and 5.24 show that the implemented method in this work is able to detect the local fault on the pinion while the accelerometer is mounted on the gear's shaft. Notice that for the case of the faulty pinion, even though the presence of the fault can be determined, still the implemented method cannot predict the angular location of the fault on pinion. This is because the number of the teeth on pinon and gear are prime, which results in a hunting tooth combination; the number of teeth of pinion and gear do not have any common prime divisor.

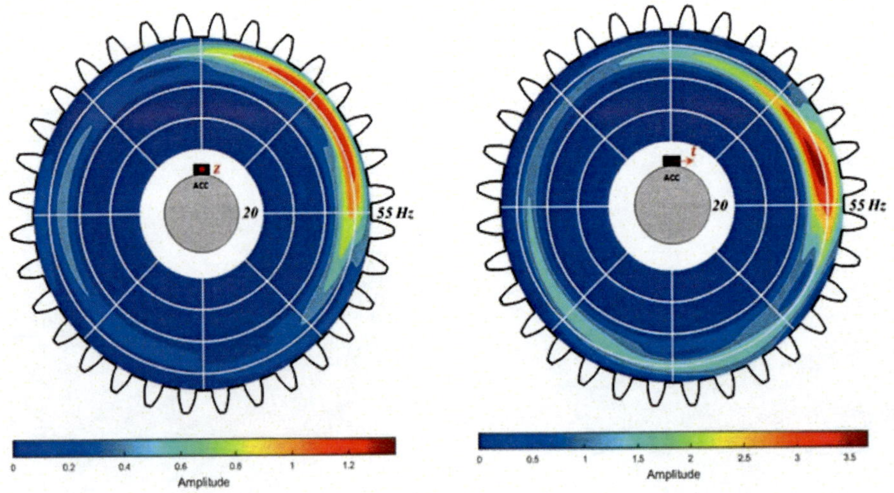

a. 25% broken tooth

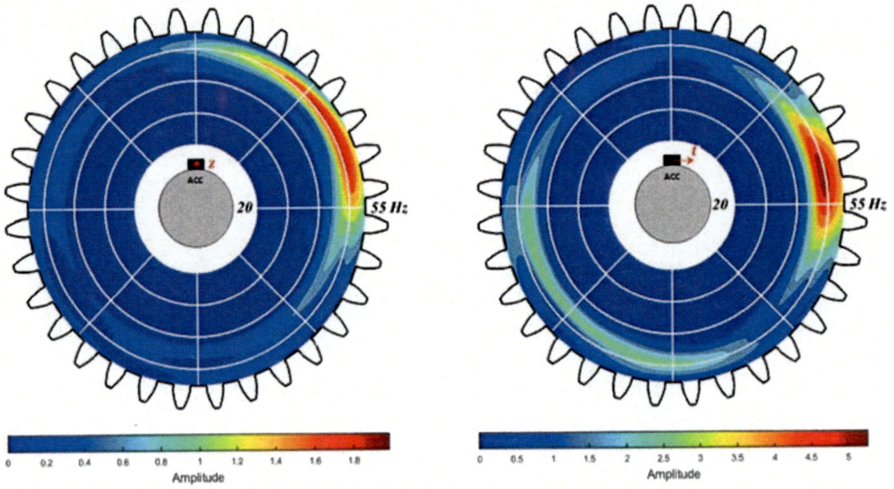

b. 50% broken tooth

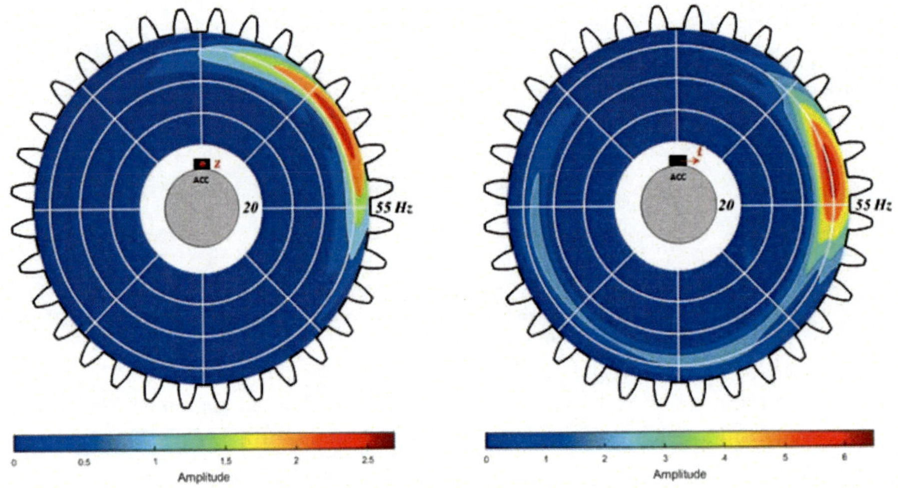

Fig. 5.20 Different size and locations for the broken tooth corresponded to Fig. 5.5 – A

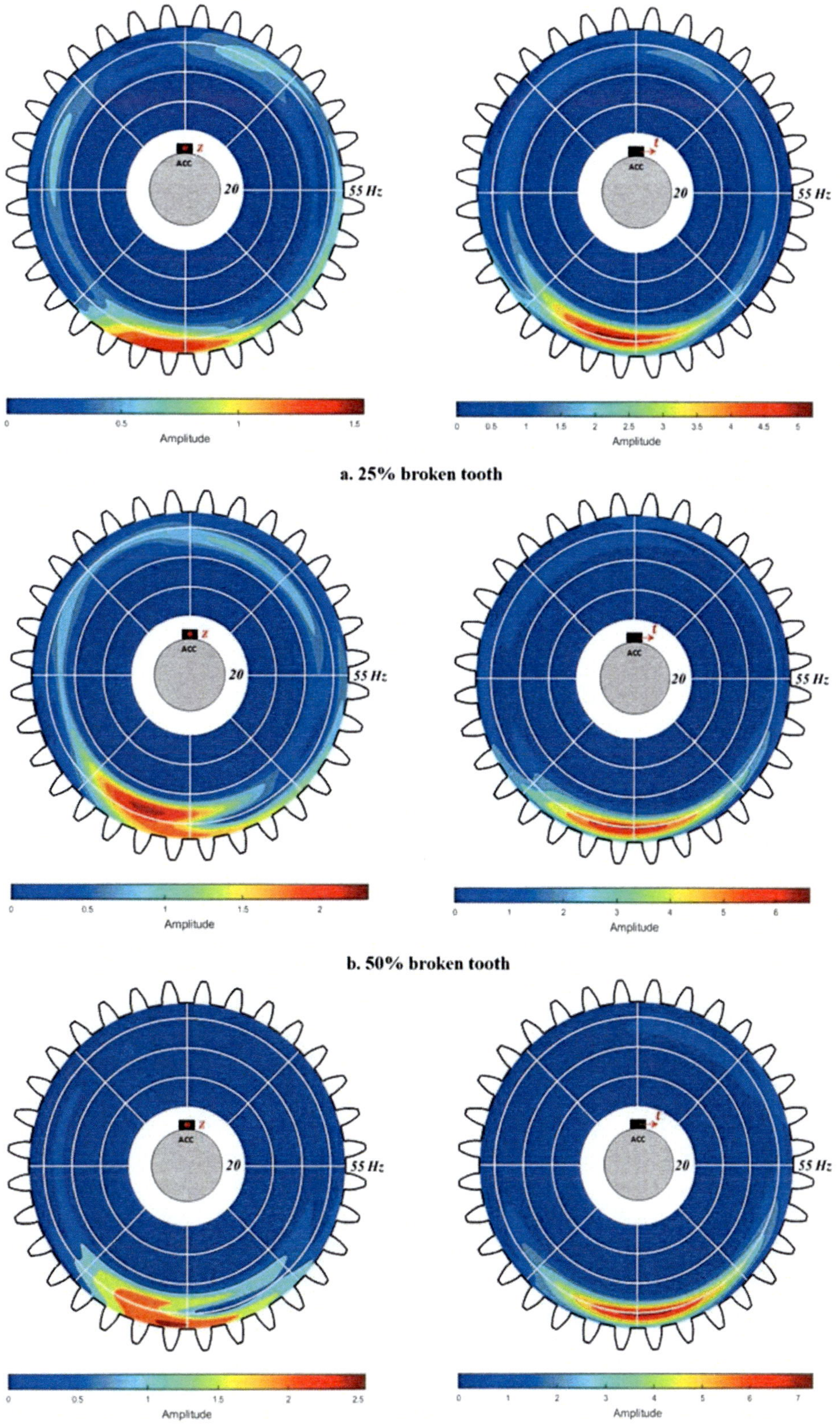

Fig. 5.21 Different size and locations for the broken tooth corresponded to Fig. 5.5 – B

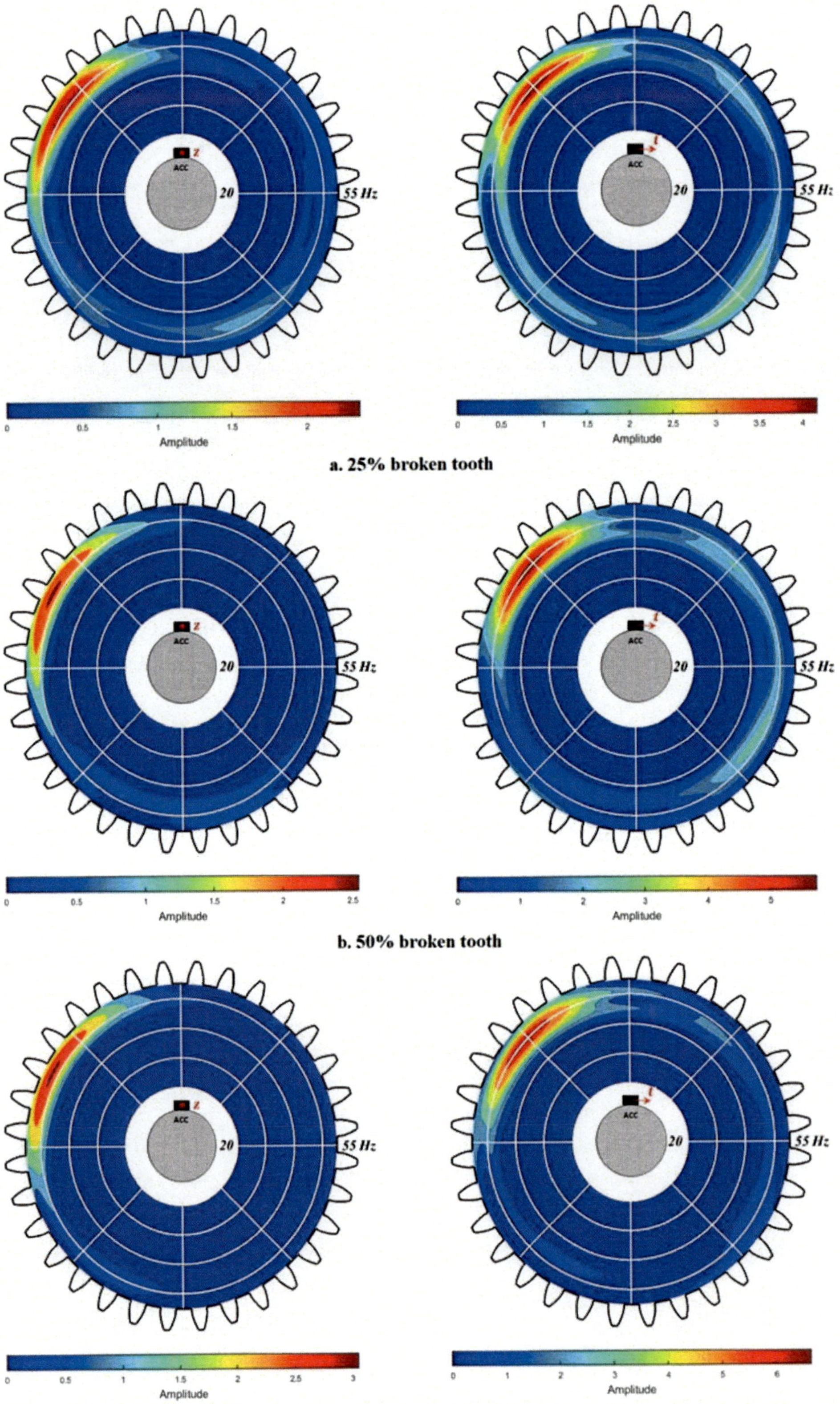

Fig. 5.22 Different size and locations for the broken tooth corresponded to Fig. 5.5 – C

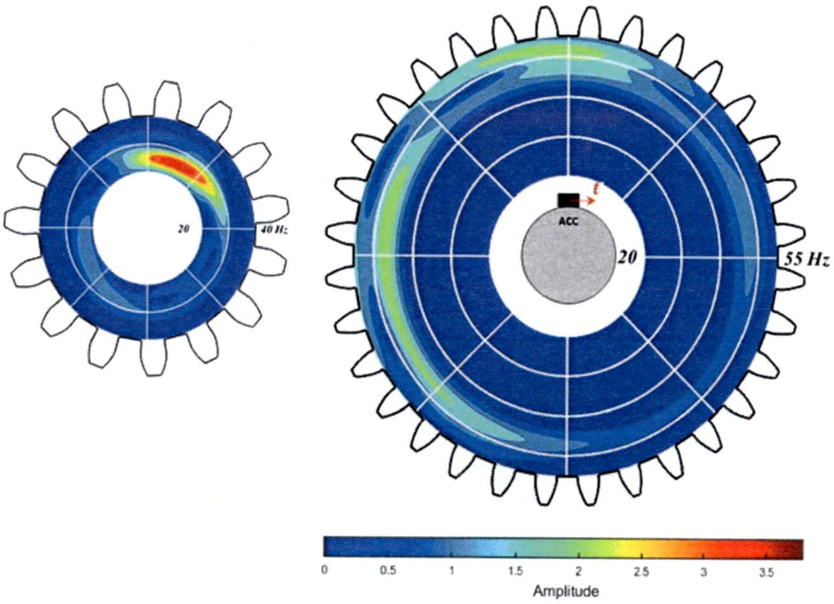

Fig. 5.23 CWT of the averaged residual signal in polar coordinate

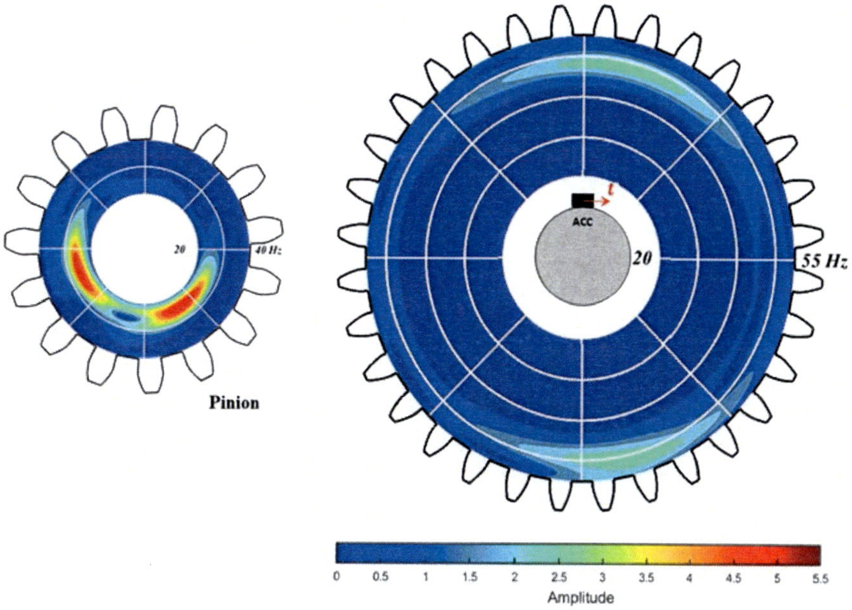

Fig. 5.24 CWT of the averaged residual signal in polar coordinate

5.14 Fluctuating Load-Speed Condition

In this section, the fault diagnostic of the healthy pinion and faulty gear, which was investigated in the previous section, is repeated under varying load-speed condition. Shaft-mounted accelerometer is used to acquire the vibration signal in tangent direction, while the rotational speed of the DC motor is fluctuated between 600 and 1000 RPM, that is, 10–17 Hz, for 2 s. Same as before, the speed of the output shaft is assumed to be constant within the two pulses of the tachometer, as demonstrated in Fig. 5.25, and the results of the signal processing are presented in Fig. 5.26. Figure 5.26 shows that the implementation of tachometer with 16 magnets in the signal acquisition and signal processing process enable us to detect and localize the faulty tooth for variable load-speeds operational condition.

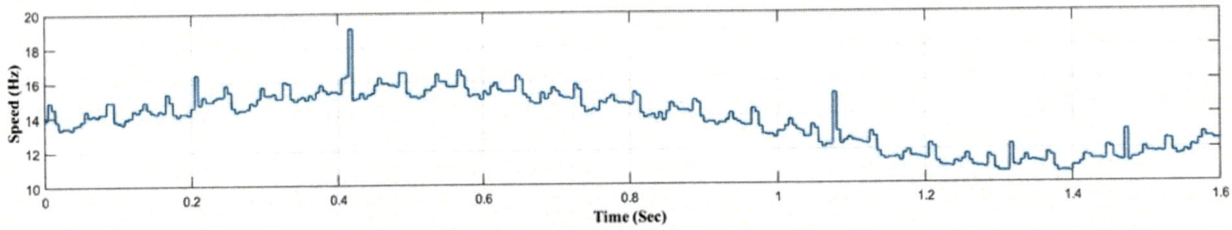

Fig. 5.25 Fluctuation speed over 20 complete revolutions

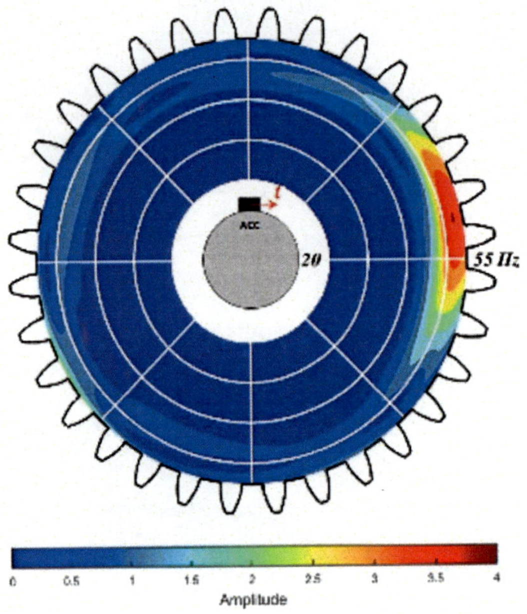

Fig. 5.26 Detection of the faulty tooth under fluctuating speed

5.15 Conclusion and Future Works

The goal of this chapter is to demonstrate the effectiveness of shaft-mounted accelerometer in the local fault diagnostics of geared systems. The results show that the shaft-mounted accelerometer is able to highlight the features of local faults in geared systems as good as the traditionally shaft-mounted accelerometer. In this work, the vibration signal measured from different locations and orientations on the shaft are compared with each other. The results show that measuring vibration signal on the surface of the shaft and in the tangent direction provides more enriched signal. The vibration signal measured on the shaft shows a good sensitivity to the degeneration and location of the fault. It is shown that as the size of the broken teeth increases, the amplitude of the short-time duration harmonic component increases. It is demonstrated that the implementation of a 16-magnets tachometer enables us to determine existence of the local fault on both pinion and gear. This is while for the prime number of pinion and gear teeth, the location of the fault can be determined only for the gear and not the pinion. At the end, it is shown that the implemented method is able to detect the fault even under fluctuating load-speed condition, which makes the implemented approach suitable for real-time health monitoring of any geared system.

5.16 Future Works

In this work, the implementation of a wireless shaft-mounted accelerometer used in the detection of local gear fault is investigated. The authors believe these results can be further expanded to enhance other objectives in fault diagnostics of rotating machinery and equipment, such as gearbox systems with multistage transmission. Other applications could also include detecting the whirling motion or eccentricity of the shafts by collecting and analyzing the data in longitudinal direction of the shaft, as well as fault detection of damaged bearings, cracked shafts, turbines with asymmetricity or angle implausibility blade, and even rotor or stator faults in a generator. Additional work could also be performed with an upgraded sensor design that is able to collect data at equal angle increments rather than equal time increment and resampling. This data acquisition system upgrade is expected to increase the quality of the collected data and reduce the computational effort and energy consumption of the circuit.

Acknowledgements This work is supported and facilitated by Ridgetop Group Inc. in Tucson, Arizona. Ridgetop is a well-established engineering and technology company that provides advanced Condition-based Maintenance (CBM), Prognostic Health Management (PHM), and reliability engineering solutions for its customers. Ridgetop's solutions aim to increase safety, efficiency, and operational performance while also reducing maintenance and sustainment costs with the most innovative products and technology for aerospace, defense, transportation, energy, and industrial applications.

References

1. Ahamed, N., Pandya, Y., Parey, A.: Spur gear tooth root crack detection using time synchronous averaging under fluctuating speed. Measurement. **52**, 1–11 (2014). https://doi.org/10.1016/j.measurement.2014.02.029
2. Hartono, D., Halim, D., Roberts, G.W., Liu, Q.: Vibration-based fault diagnostic of a spur gearbox. MATEC Web Conf. **70**, 1–5 (2016). https://doi.org/10.1051/matecconf/20167002004
3. Hartono, D., Halim, D., Widodo, A., Roberts, G.W.: Time-frequency enhancement technique for bevel gear fault diagnosis. MATEC Web Conf. **70** (2016). https://doi.org/10.1051/matecconf/20167002003
4. Buzzoni, M., Mucchi, E., D'Elia, G., Dalpiaz, G.: Diagnosis of localized faults in multistage gearboxes: a vibrational approach by means of automatic EMD-based algorithm. Shock. Vib. **2017** (2017). https://doi.org/10.1155/2017/8345704
5. Sharma, V., Parey, A.: Frequency domain averaging based experimental evaluation of gear fault without tachometer for fluctuating speed conditions. Mech. Syst. Signal Process. **85**, 278–295 (2017). https://doi.org/10.1016/j.ymssp.2016.08.015
6. Li, H., Zhang, Y., Zheng, H.: Angle domain average and CWT for fault detection of gear crack. In: Proceedings of the 5th International Conference on Fuzzy Systems and Knowledge Discovery FSKD 2008, vol. 3, pp. 137–141 (2008). https://doi.org/10.1109/FSKD.2008.62
7. Heidari Bafroui, H., Ohadi, A.: Application of wavelet energy and Shannon entropy for feature extraction in gearbox fault detection under varying speed conditions. Neurocomputing. **133**, 437–445 (2014). https://doi.org/10.1016/j.neucom.2013.12.018
8. Bonnardot, F., El Badaoui, M., Randall, R.B., Danière, J., Guillet, F.: Use of the acceleration signal of a gearbox in order to perform angular resampling (with limited speed fluctuation). Mech. Syst. Signal Process. **19**, 766–785 (2005). https://doi.org/10.1016/j.ymssp.2004.05.001
9. Fedala, S., Rémond, D., Zegadi, R., Felkaoui, A.: Contribution of angular measurements to intelligent gear faults diagnosis. J. Intell. Manuf. **29**, 1115–1131 (2018). https://doi.org/10.1007/s10845-015-1162-1
10. Fyfe, K.R., Munck, E.D.S.: Analysis of computed order tracking. Mech. Syst. Signal Process. **11**, 187–205 (1997). https://doi.org/10.1006/mssp.1996.0056
11. Singh, A., Parey, A.: Gearbox fault diagnosis under non-stationary conditions with independent angular re-sampling technique applied to vibration and sound emission signals. Appl. Acoust. **144**, 11–22 (2019). https://doi.org/10.1016/j.apacoust.2017.04.015
12. Parey, A., Singh, A.: Gearbox fault diagnosis using acoustic signals, continuous wavelet transform and adaptive neuro-fuzzy inference system. Appl. Acoust. **147**, 133–140 (2019). https://doi.org/10.1016/j.apacoust.2018.10.013
13. Sharma, V., Parey, A.: Gear crack detection using modified TSA and proposed fault indicators for fluctuating speed conditions. Meas. J. Int. Meas. Confed. **90**, 560–575 (2016). https://doi.org/10.1016/j.measurement.2016.04.076
14. Li, Y., Ding, K., He, G., Lin, H.: Vibration mechanisms of spur gear pair in healthy and fault states. Mech. Syst. Signal Process. **81**, 183–201 (2016). https://doi.org/10.1016/j.ymssp.2016.03.014
15. Shivputra, A.S., Shrigandhi, G.D.: Gear fault identification by using vibration analysis. Int. J. Curr. Eng. Technol. **4**, 247–253 (2011). https://doi.org/10.14741/ijcet/22774106/spl.4.2016.51
16. Ma, J., Yang, X., Wang, D., Shao, W.: Research on vibration signal analysis and extraction method of gear local fault. Proc. SPIE. **66** (2018). https://doi.org/10.1117/12.2307685
17. Bendjama, H., Bouhouche, S., Boucherit, M.S.: Application of wavelet transform for fault diagnosis in rotating machinery. Int. J. Mach. Learn. Comput. **2**, 82–87 (2012). https://doi.org/10.7763/ijmlc.2012.v2.93
18. Zhang, X., Kang, J., Bechhoefer, E., Zhao, J.: A new feature extraction method for gear fault diagnosis and prognosis. Eksploat. i Niezawodn. **16**, 295–300 (2014)
19. Wilk, A., Łazarz, B., Madej, H.: Gear fault detection using vibration analysis. Diagnostyka. nr. **3**(47), 111–116 (2008)
20. Dalpiaz, G., Rivola, A., Rubini, R.: Gear fault monitoring: comparison of vibration analysis techniques. In: Proceedings of the 3rd International Conference Acoustical and Vibratory Surveillance Methods and Diagnostics Techniques, pp. 623–632 (1998). http://diem1.ing.unibo.it/mechmach/rivola/pub13.pdf
21. Ding, Y., He, W., Chen, B., Zi, Y., Selesnick, I.W.: Detection of faults in rotating machinery using periodic time-frequency sparsity. J. Sound Vib. **382**, 357–378 (2016). https://doi.org/10.1016/j.jsv.2016.07.004
22. Tse, P.W., Yang, W.X., Tam, H.Y.: Machine fault diagnosis through an effective exact wavelet analysis. J. Sound Vib. **277**, 1005–1024 (2004). https://doi.org/10.1016/j.jsv.2003.09.031
23. Jafarizadeh, M.A., Hassannejad, R., Ettefagh, M.M., Chitsaz, S.: Asynchronous input gear damage diagnosis using time averaging and wavelet filtering. Mech. Syst. Signal Process. **22**, 172–201 (2008). https://doi.org/10.1016/j.ymssp.2007.06.006
24. Lin, J., Zuo, M.J.: Gearbox fault diagnosis using adaptive wavelet filter. Mech. Syst. Signal Process. **17**, 1259–1269 (2003). https://doi.org/10.1006/mssp.2002.1507

25. Lokesha, M., Majumder, M., Ramachandran, K., Raheem, K.: Fault diagnosis in gear using wavelet envelope power spectrum. Int. J. Eng. Sci. Technol. **3**, 156–167 (1970). https://doi.org/10.4314/ijest.v3i8.13
26. Han, D., Li, P., An, S., Shi, P.: Multi-frequency weak signal detection based on wavelet transform and parameter compensation band-pass multi-stable stochastic resonance. Mech. Syst. Signal Process. **70–71**, 995–1010 (2016). https://doi.org/10.1016/j.ymssp.2015.09.003
27. Meltzer, G., Dien, N.P.: Fault diagnosis in gears operating under non-stationary rotational speed using polar wavelet amplitude maps. Mech. Syst. Signal Process. **18**, 985–992 (2004). https://doi.org/10.1016/j.ymssp.2004.01.009
28. D.G. (NASA G.R.C. Lewicki Cleveland, OH United States), N.A. (Ridgetop G. Lambert Inc., Tucson, AZ, United States), R.S. (Ridgetop G. Wagoner Inc., Tucson, AZ, United States), N.T.R. Server(NTRS), Evaluation of MEMS-Based Wireless Accelerometer Sensors in Detecting Gear Tooth Faults in Helicopter Transmissions, 40. http://ridgetopgroup.actonsoftware.com/acton/ct/2522/s-0182-1511/Bct/l-1e22/l-1e22:5/ct2_0/1?sid=U64wPipoQ (2015)
29. Elnady, M.E., Sinha, J.K., Oyadiji, S.O.: Condition Monitoring of Rotating Machines Using On-Shaft Vibration Measurement. Woodhead Publishing Limited (2012). https://doi.org/10.1533/9780857094537.10.669
30. Hartono, D., Halim, D., Roberts, G.W.: Gear fault diagnosis using the general linear chirplet transform with vibration and acoustic measurements. J. Low Freq. Noise Vib. Act. Control. **38**, 36–52 (2019). https://doi.org/10.1177/1461348418811717
31. Liang, L., Liu, F., Kong, X., Li, M., Xu, G.: Application of instantaneous rotational speed to detect gearbox faults based on double encoders. Chin. J. Mech. Eng. **32**, 9 (2019). https://doi.org/10.1186/s10033-019-0324-z
32. Chen, J., Li, Z., Pan, J., Chen, G., Zi, Y., Yuan, J., Chen, B., He, Z.: Wavelet transform based on inner product in fault diagnosis of rotating machinery: A review. Mech. Syst. Signal Process. **70–71**, 1–35 (2016). https://doi.org/10.1016/j.ymssp.2015.08.023

Chapter 6
Methodologies to Distinguish Locomotive Types Based on the Measured Geometry

Mariia Zakharenko, Gunnstein T. Frøseth, and Anders Rönnquist

Abstract This chapter presents methods to sort locomotives and other certain types of vehicles by types using geometry measured by a measurement station. The approach is implemented using a series of artificial locomotives created by adding normally distributed noise to each distance between adjustment wheels of locomotives that are in exploitation in Norway. Two approaches are studied. The first approach, Gaussian Mixture Clustering, is presented and shown to be suitable to obtain geometries of the main locomotives in the set. The second approach, based on the comparison of norms of differences between geometries of measured and known locomotives, is found quite accurate to use for sorting measured locomotives by the predefined types. The methods are presented on geometries of locomotives, but they also can be used for distinguishing different types of wagons and multiple units.

Keywords Steel railway bridges · Locomotives · Strain measurements · Gaussian mixture clustering · Euclidean norm

6.1 Introduction

Fatigue is a primary damage mechanism in steel railway bridges. To estimate the remaining fatigue life of such structures, load models are used, for example [1–3]. Fatigue load model is a representation of traffic conditions that are used to estimate the fatigue damage in a bridge. However, models request calibration coefficients to represent actual traffic conditions that are obtained using measurements of trains passing a bridge.

A cheap and convenient measurement station was presented in [4]. The station measures strains in six strain sensors reflecting strains induced in rails by each wheel of the passing train. The time lap between the peaks corresponding to adjacent wheels and measured in the same channel shows the distance between wheels and makes it possible to obtain the geometry (distances between all wheels) of the train passed the measurement station. However, the raw data do not contain information about types of vehicles forming the train. Moreover, according to [4], calibration of sensors is performed using locomotives with known geometries and passed the measurement station. In case if geometries of locomotives in train traffic are unknown (there is no database of locomotive's geometries), obtaining calibration coefficients for the sensor becomes extremely difficult. This process requires performing field tests using real trains to calibrate the measurement station, which can be expensive.

The current study suggests methodologies to use machine learning approaches in clustering to define the database consisting of main locomotives that were measured. The methodology is presented using locomotives, but it can be also used to distinguish between the wagon types. In addition, the study examines the approach that was used in [4] to filter the known locomotives according to the types.

6.2 Methods

A series of artificial locomotives was generated based on the database of locomotives that have been used in the Norwegian railway network [5] for more than the last 20 years. The locomotives database from [5] consists of two types of locomotives

M. Zakharenko (✉) · G. T. Frøseth · A. Rönnquist
Department of Structural Engineering, Norwegian University of Science and Technology (NTNU), Trondheim, Norway
e-mail: maria.zakharova@ntnu.no; gunnstein.t.froseth@ntnu.no; anders.ronnquist@ntnu.no

C. Walber et al. (eds.), *Sensors & Instrumentation and Aircraft/Aerospace Testing Techniques, Volume 8*, Conference Proceedings of the Society for Experimental Mechanics Series, https://doi.org/10.1007/978-3-031-34938-6_6

Table 6.1 Database of locomotives that are in the service in the Norwegian railway network

	Length, m	Pitch1, m	Pitch2, m	Pitch3, m	Pitch4, m	Pitch5, m
Bo′Bo′-a	10.4	3	4.4	3	–	–
Bo′Bo′-b	10.5	3.2	4.1	3.2	–	–
Bo′Bo′-c	13.8	2.8	8.2	2.8	–	–
Bo′Bo′-d	11.4	2.4	6.6	2.4	–	–
Bo′Bo′-e	10.4	2.7	5	2.7	–	–
Bo′Bo′-f	13	2.6	7.8	2.6	–	–
Co′Co′-a	14.3	2	2	6.3	2	2
Co′Co′-b	12	1.8	1.8	4.8	1.8	1.8
Co′Co′-c	15.7	1.8	2.1	7.9	2.1	1.8
Co′Co′-d	18.2	1.8	1.8	11	1.8	1.8
Co′Co′-e	17.2	2	2.1	9	2.1	2
Co′Co′-f	15.5	1.8	2	7.9	2	1.8

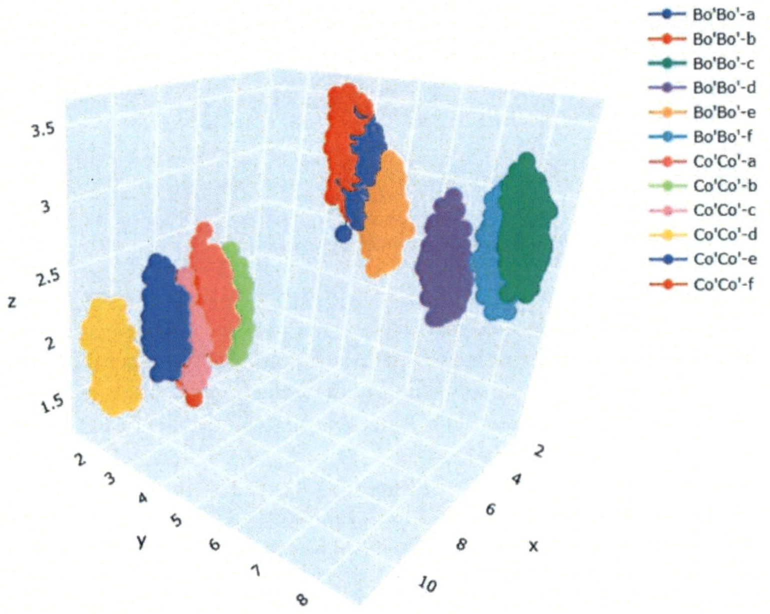

Fig. 6.1 The features of samples of generated locomotives. OX is third pitch, OY is second pitch, OZ is first pitch

(six subtypes of four-axle locomotives and six subtypes of six-axle locomotives, see Table 6.1). The geometry of the locomotives was generated by adding a normally distributed noise term ($u = 0$, $s = 0.1$) to each pitch (distance between adjacent wheels) of the database locomotive since the measurement error is normally distributed.

Figure 6.1 presents the illustration of the first three pitches for each sample. There are two problematic areas having overlapping distributions of features: fully overlapped Co′Co′-c, Co′Co′-f; partly overlapped Bo′Bo′-a, Bo′Bo′-b, Bo′Bo′-e. The area of Bo′Bo′-c and Bo′Bo′-f is not overlapping, but the type of analysis should be chosen accurately due to the closeness of features of these two groups.

The focus of the study is two methods having different goals. The first method, Gaussian Mixture Clustering, has a goal to obtain the main groups of vehicles and their geometries. The first approach is supposed to be used to obtain the locomotives database, as in Table 6.1, for cases where there is no predefined database, but the measured data exists. The method is described in [6], and the assumption in the base of the method is that the data points follow Gaussian distribution, which fits the case of the study since such a distribution is the expected error of the geometry. An Expectation-Maximization form of optimization function is used in the method to find the mean and variance. First, the random parameters of the distribution are chosen, and the sample is checked if it belongs to the cluster. Second, maximization updates the Gaussian parameters to fit the points assigned to the cluster. The aim of the second step is to increase the likelihood of the sample belonging to the cluster distribution.

The Norm method is aimed to distinguish vehicles into groups using a predefined database of geometries that can be obtained with the help of the first method. It was presented in [4] and used by choosing the minimum norm of the difference between the database locomotive geometry and measured geometry, which pointed out the most similar geometry. The Euclidian norm is the distance from the origin to the vector coordinates.

6.3 Results

The results of the application of both methods are presented in Fig. 6.2. The Gaussian Mixture Clustering was performed with parameter tol = 0.001. The parameter tol defines convergence threshold: iterations stop when the lower bound average gain is below this threshold. The distribution of norms in the Norm method is presented in Fig. 6.3.

Both methods are not 100% perfect in the task to distinguish overlapping areas. The coefficient that is used to reflect the measure of clustering quality is Rand Index [7], see formula (6.1):

$$RI = \frac{TP + TN}{TP + FP + FN + TN} \tag{6.1}$$

where TP is the number of true-positive pairs (the number of pairs that initially were at the same group and end up at the same cluster), TN is the number of true-negative pairs, FP is the number of false-positive pairs, and FN is the number of false-negative pairs. The Rand Index of Gaussian Mixture Clustering is 0.9757, the Norm approach has Rand Index equal to 0.9805. The Norm approach was used in [4] while distinguishing the types of locomotives. In the current study, the approach showed good quality and was useful.

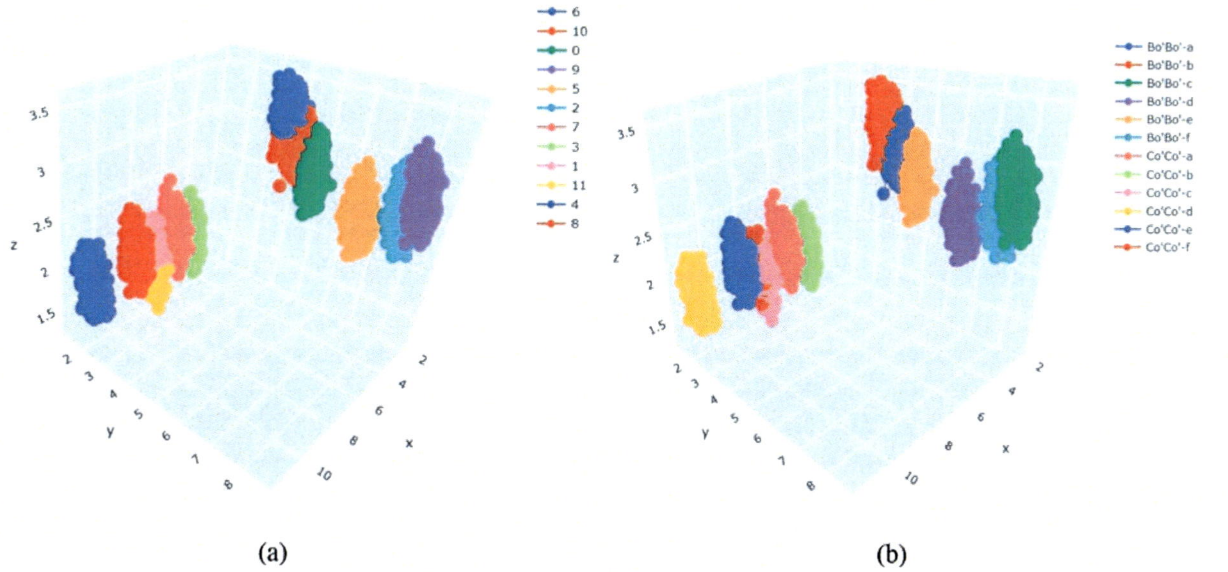

Fig. 6.2 Results of (**a**) Gaussian Mixture Clustering, (**b**) the Norm approach

Fig. 6.3 Distribution of norms of locomotives

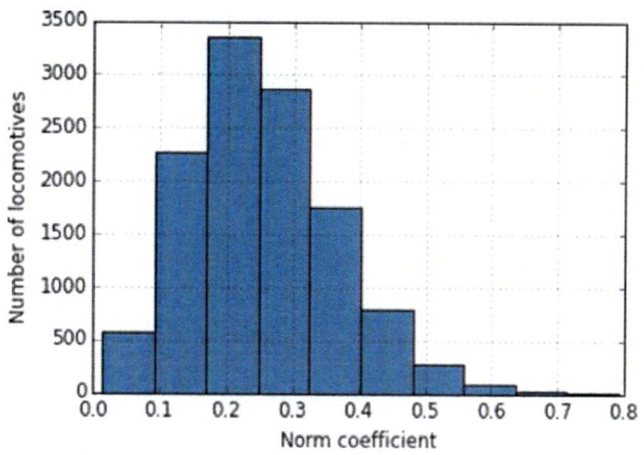

Table 6.2 Geometries of locomotives obtained from the results of (a) Gaussian Mixture Clustering, (b) the Norm approach. There are presented mean values of features

(a) Gaussian mixture clustering

	p1, m	p2, m	p3, m	p4, m	p5, m
0	2.686905	5.01095	2.698049	0	0
1	1.912148	1.995356	7.900456	2.045339	1.795411
2	2.604097	7.800223	2.595561	0	0
3	1.796878	1.798752	4.807267	1.793646	1.80415
4	1.806341	1.800113	10.99972	1.799057	1.796648
5	2.395502	6.600808	2.401568	0	0
6	3.241016	4.102206	3.172676	0	0
7	1.99501	2.010399	6.301617	1.991309	2.00197
8	2.000495	2.101389	9.00886	2.087942	1.999084
9	2.794946	8.204215	2.798526	0	0
10	2.966545	4.397411	3.0179	0	0
11	1.689519	2.107552	7.905357	2.042511	1.798595

(b) The norm approach

	p1, m	p2, m	p3, m	p4, m	p5, m
Bo′Bo′-a	2.994472	4.412938	2.987052	0	0
Bo′Bo′-b	3.215679	4.078167	3.212175	0	0
Bo′Bo′-c	2.79475	8.204687	2.798439	0	0
Bo′Bo′-d	2.395502	6.600808	2.401568	0	0
Bo′Bo′-e	2.690578	5.002751	2.699045	0	0
Bo′Bo′-f	2.604484	7.800155	2.595851	0	0
Co′Co′-a	1.99501	2.010399	6.301617	1.991309	2.00197
Co′Co′-b	1.796878	1.798752	4.807267	1.793646	1.80415
Co′Co′-c	1.757925	2.14721	7.830596	2.135017	1.759834
Co′Co′-d	1.806341	1.800113	10.99972	1.799057	1.796648
Co′Co′-e	2.000495	2.101389	9.00886	2.087942	1.999084
Co′Co′-f	1.83495	1.965694	7.969911	1.959601	1.831461

The initial idea of the application of the methods is next. The Clustering method is used to create a database of vehicles in raw data, and it is shown to be suitable for the purpose. The Norm approach is more sensitive to overlapping areas, even though it may not achieve 100% quality, but the Rand Index, on the other hand, remains high. Table 6.2 presents tables containing geometries of locomotives' groups that were obtained using both methods. The table consists of mean values of parameters among all locomotives in the group.

The database obtained using the first method is very similar to the original base, see Table 6.1, as well as the database defined using the second approach. The number of locomotives in both groups is about 1000 as it was in the initial database. Locomotive types that do not have overlapping in features distributions, Bo′Bo′-c, Bo′Bo′-d, Bo′Bo′-f, Co′Co′-a, Co′Co′-b, Co′Co′-d, Co′Co′-e, have 100% quality of clustering; the remaining groups also have a high percentage. That means both approaches are useful for distinguishing the different types of locomotives.

6.4 Conclusion

The goal of the study is to present and test approaches to obtain locomotive types in the raw data containing geometries of locomotives without marks and to sort raw data by preordered list of locomotives.

With the goal to test both approaches 12.000 locomotives have been created based on the database of Norwegian locomotives of the current time, Gaussian distribution was used to simulate measurement error in geometry.

The clustering called Gaussian Mixture is admittedly the most efficient in terms of obtaining groups of locomotives in unmarked set of locomotives. We chose Gaussian Mixture because the measurement error follows a normal distribution, assuming that the clusters also have Gaussian distributions of features. Several approaches, such as K-Means, DBSCAN,

Hierarchical Clustering, have been studied in addition to the Gaussian Mixture; however, they did not show satisfactory quality and were not included in the report. The quality of Gaussian Mixture is calculated and equals 0.9757, which shows the approach is appropriate to use in forming the database of geometry of the main types of locomotives in a set of measured locomotives. The quality of the method can be used by taking into account the measured loads.

The Norm approach was introduced earlier in [4]. The current study verifies the quality of the approach using created artificial locomotives. The Rand Index of the approach is higher than in the case of the first method and equals 0.9805. The Norm approach is shown to be useful and accurate to divide locomotives according to the database of the types of locomotives.

The current approaches are tested on the locomotives. However, the promising direction of study is to use Gaussian Mixture Clustering for distinguishing the main groups of wagons since the variation of such vehicles is much higher than variation of locomotives, and it is very useful to have marks of the wagons.

References

1. Frøseth, G.T., Rönnquist, A.: Load model of historic traffic for fatigue life estimation of Norwegian railway bridges. Eng. Struct. **200**, 109626 (2019)
2. Imam, B., Salter, P.A.: Historical load effects on fatigue of metallic railway bridges. In: Proceedings of the Institution of Civil Engineers-Bridge Engineering, vol. 171, pp. 49–62. Thomas Telford Ltd (2018)
3. Hulet, K.M., Smith, C.C., Gilbert, M.: Load-carrying capacity of flooded masonry arch bridges. In: Proceedings of the Institution of Civil Engineers-Bridge Engineering, vol. 159, pp. 97–103. Thomas Telford Ltd (2006, September)
4. Zakharenko, M., Frøseth, G.T., Rönnquist, A.: Train classification using a weigh-in-motion system and associated algorithms to determine fatigue loads. Sensors. **22**(5), 1772 (2022)
5. Frøseth, G.T., Rönnquist, A.: Evolution of load conditions in the Norwegian railway network and imprecision of historic railway load data. Struct. Infrastruct. Eng. **15**(2), 152–169 (2019)
6. Zeng, H., Cheung, Y.M.: A new feature selection method for Gaussian mixture clustering. Pattern Recogn. **42**(2), 243–250 (2009)
7. Yeung, K.Y., Ruzzo, W.L.: Details of the adjusted rand index and clustering algorithms, supplement to the paper an empirical study on principal component analysis for clustering gene expression data. Bioinformatics. **17**(9), 763–774 (2001)

Chapter 7
Feedback Active Noise Cancellation Using Single Sensor with Deep Learning

Alireza Mostafavi and Young-Jin Cha

Abstract Constructive measures should be taken immediately to tackle urban noise pollution, which is an omnipresent but neglected threat to human health. Many attempts have been made by researchers during recent decades to alleviate this issue; however, due to the nature of linear filters, conventional active noise control (ANC) methods, for example, filtered-x least mean square (FxLMS) algorithm, are useful just for attenuating narrowband linear or tonal noises. To deal with environmentally complex ANC applications, we developed a new deep learning-based artificial intelligence algorithm, which is able to model the intrinsic nonlinear behavior of various noises and produce anti-noise, which destructively interferes with unwanted noise to neutralize it. The proposed algorithm as a feedback controller significantly outperformed the traditional feedback FxLMS method in terms of noise attenuation metric.

Keywords Active noise cancellation · Deep learning · Primary and secondary paths · Feedback control · Vehicle noise

7.1 Introduction

Due to the large-scale developments and urbanizations, a growing number of people are subjected to excessive noise pollution. Statistics show that over two billion people around the world are exposed to environmental, road traffic, neighbors, and bar noises levels over 55 dB by 2011 [1]. Exposure to this high level of noise for a long period can cause physical and mental issues [2]. So, worthy measures should be taken immediately to control this situation and decrease noise levels in densely populated areas.

Noise cancellations or mitigations can be categorized into two main groups: passive noise cancellation (PNC) or active noise cancellation/control (ANC). PNC generally refers to the isolation of a space from a noisy environment by implementing sound and vibration absorbers equipment, such as soundproofing foam, dampers, barriers, and acoustic fabric [3]. This is the most traditional way of decreasing noise levels in a specific enclosed environment. Although these passive methods do not require power, they mostly impede airflow, and more importantly, they have negligible effects on low-frequency noises (i.e., less than 1000 Hz) [4]. These shortcomings of PNC techniques have shifted scientists' attention to ANC methods.

In general, ANC can be defined as producing canceling signals or anti-noise signals with the same amplitude and opposite phase of undesired noise to neutralize it [5]. As the name suggests, it is an active method that requires some basic active equipment. A typical ANC system consists of a microphone to receive noise from noisy environments, which is called reference microphone, a loudspeaker to generate anti-noise signals, an error microphone to measure the residual noise, and other acoustic devices to make this equipment work, such as amplifiers, analog to digital converter (ADC), and digital to analog converters (DAC). This ANC system that works with two microphones is known as feed-forward (FF) ANC. There is another ANC technique called feedback (FB) ANC, which only requires an error microphone and loudspeaker. Both systems aim to cancel noise levels at error microphone locations that should be installed in the region of interest (ROI), which we want to be a quiet place. Furthermore, they have different applications based on their basic requirements. On the one hand, feed-forward ANC is mostly preferred where the noise sources are stationary with respect to the ROI. This can help the FF method to track the reference noise easily. For instance, FF has usually been used in aircraft [6] and automobile [7] cabins to cancel out engine noise in the passengers' area. On the other hand, the FB ANC system is preferred when dealing with

A. Mostafavi (✉) · Y.-J. Cha
Department of Civil Engineering, University of Manitoba, Winnipeg, MB, Canada
e-mail: mostafa3@myumanitoba.ca; young.cha@umanitoba.ca

C. Walber et al. (eds.), *Sensors & Instrumentation and Aircraft/Aerospace Testing Techniques, Volume 8*, Conference Proceedings of the Society for Experimental Mechanics Series, https://doi.org/10.1007/978-3-031-34938-6_7

moving noise sources such as construction equipment and workers on construction sites. As a result, this feature of FB ANC makes it more versatile than FF ANC.

More important than physical equipment, we need an algorithm to accurately estimate compensation signal or anti-noise, which destructively superposes with ambient noise. Filtered x least mean square (FxLMS) [5] and its variations are the most broadly utilized ANC algorithms due to their low cost of computations and robustness [8]. The FxLMS is developed to account for the phase distortion and nonlinear effect of secondary acoustic and electrical paths from the loudspeaker to the error microphone location [5]. Moreover, it is an adaptive algorithm that aims to adapt two finite impulse response (FIR) filters for modeling secondary path and anti-noise estimation. Owing to the linear nature of FIR filters, the FxLMS fails to efficiently model the true secondary path and produce canceling signals in real-world scenarios with existing nonlinearities in noises and acoustic devices.

Many attempts have been made to tackle this problem and improve the performance of the FxLMS method [9, 10]. However, the performance of traditional methods and their extensions are restricted when there are strong nonlinearities in acoustic devices or in unwanted noise. In recent years, researchers have incorporated machine learning (ML) and artificial intelligence algorithms to address nonlinear ANC. In [11], a three-layer multilayer perceptron artificial neural network (ANN) is proposed to attenuate broadband noises. Na and Chae [12] used a recurrent neural network (RNN) for feedback single sensor ANC. Through simulations, they have shown that RNN performs better than ANN and FxLMS algorithms when dealing with nonstationary and time-varying noises. Park et al. [13] developed two networks using long-short term memory (LSTM) and convolutional neural networks (CNN) for FF ANC, and they could get better results than ANN and RNN models. Nevertheless, all of these studies did not consider primary and secondary path effects in their ANC system, and they oversimplified the ANC problem. The primary path is the acoustic path between the reference microphone to the error microphone location, which introduces a delay and phase shifts when noise waves propagate between them. They might reach excellent results in simulations; however, in practical applications of ANC, there are unavoidable nonlinearities in primary and secondary paths, which significantly reduce the effectiveness of methods that are not able to precisely model ANC nonlinear distortions.

Recent advancements in the deep learning-based artificial neural network [14–21], and access to a huge amount of data have significantly changed our world, and the active noise cancellation area is no exception. Zhang & Wang [22] have used a speech enhancement model for FF frequency domain ANC. Considering all nonlinearities in the ANC system, their deep model could get good noise attenuation results in both noisy and, more specifically, noisy speech environments, which all conventional algorithms have serious limitations. However, the network they have utilized is pretty huge with several millions of parameters that were primarily designed for speech enhancement. In the speech enhancement research area, there is no severe restriction for latency. However, in FF ANC systems, all processing time and anti-noise propagation time from loudspeaker to error microphone should be less than the time it takes for primary noise to travel from reference microphone to error microphone location in order to respect causality constraint. Otherwise, the ANC system would have zero or negligible effect. Although they have predicted upcoming anti-noise samples up to 20 milliseconds, they still need powerful graphical processing units (GPUs) to make their network work in real-time by considering causality constraint.

In this chapter, we have proposed a deep leaning-based model that can be used as a feedback controller for ANC applications. It is a lightweight network with just 345 parameters that can be easily implemented using GPU computations or even microcontroller units which have been recently utilized to implement deep learning models [23]. The developed algorithm is able to predict anti-noise signals in a real-time manner with consideration of primary and secondary path nonlinearities. In the next sections, the methodology of the FB ANC system is described, followed by the simulation procedure. Then the network design along with simulation data is explained. Subsequently, the results are presented, and a conclusion is drawn in the last section.

7.2 Proposed Single Sensor ANC Framework

Figure 7.1 illustrates the overall proposed ANC framework where $x(n)$ is the primary noise at elapse time n, $d(n)$ is desired noise that should be canceled, $y(n)$ is the signal generated by ANC algorithm which passes through secondary path $S(z)$ to make anti-noise signals. In FB ANC, since we do not have access to the desired noise signal $d(n)$, we have to regenerate it internally by summing the error signal $e(n)$ and generated anti-noise by ANC y (n) at previous elapse time. The residual noise can be calculated using Eq. 7.1 as follows:

$$e(n) = d(n) - a(n) \tag{7.1}$$

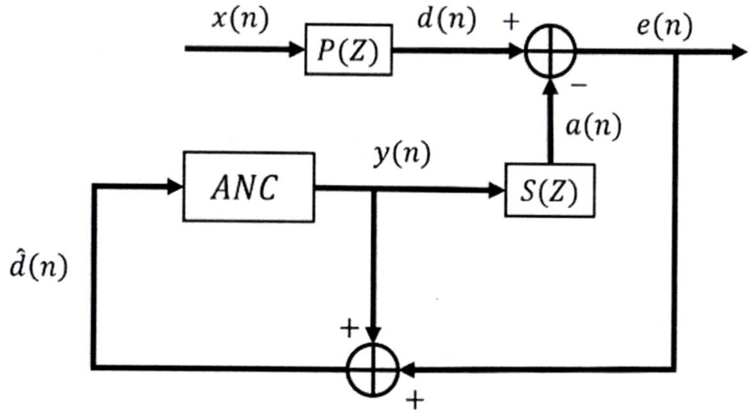

Fig. 7.1 block diagram of proposed single sensor-based FB ANC

Where $d(n)$ and $a(n)$ are defined by Eqs. 7.2 and 7.3, respectively.

$$d(n) = x(n) * p(n) \tag{7.2}$$

$$a(n) = ANC\left(\hat{d}(n)\right) * (n) \tag{7.3}$$

The * operation is linear convolution. In relation (7.3) the $\hat{d}(n)$ is computed by the following equation:

$$\hat{d}(n) = y(n-1) + e(n) \tag{7.4}$$

The input to the ANC network in Fig. 7.1 is a vector with 20 samples, and the output is the one sample prediction of anti-noise. Meaning that in every time step, the input to the ANC network is updated by one sample of estimated desired noise and fed to the ANC network for prediction of compensation signal in the next time step.

7.3 Primary and Secondary Paths Simulations

In Fig. 7.1, $P(z)$ and $S(Z)$ are primary and secondary paths which are usually simulated by the room impulse response (RIR) with a finite duration, which is called finite impulse response filter (FIR) [24, 25]. In this study, we have simulated a rectangular room with [3 m, 4 m, 2 m] dimensions and used the library developed in [24], which is based on the image theory described in [25] to generate RIRs for $P(z)$ and $S(z)$. Error microphone location is set at [1.5 m, 2.5 m, 1 m], canceling loudspeaker at [1.5 m, 3 m, 1 m] and noise source at [1.5 m, 1 m, 1 m]. The length of RIRs is chosen 1500 samples, and the room's reverberation time (RT60) is set at 0.2 s, which is an average RT60 for a room of this size [22]. In FB ANC, it is essential that the distance between the noise source and error microphone should be greater than the distance between the loudspeaker and the error microphone; otherwise, the causality constraint would not be respected even for zero processing delay. So, we assumed the noise source is three times further than the anti-noise source to the error microphone location in order to secure enough time for the ANC algorithm to produce anti-noise.

7.4 ANC Network Architecture

Figure 7.2 shows the proposed deep learning ANC algorithm, which is designed by carefully integrating traditional 1D convolution (Conv), 1D atrous convolution (Aconv), recurrent neural network (RNN) of type of Elman RNN [26], fully connected layer (FC), and nonlinear activation functions such as scaled exponential linear unit (SeLU) and hyperbolic tangent (Tanh) in three consecutive layers. Aconv with various dilation factors and number of filters are used to extract multilevel features from different receptive fields of raw time-domain signal, followed by SeLU to model intrinsic nonlinearities in the

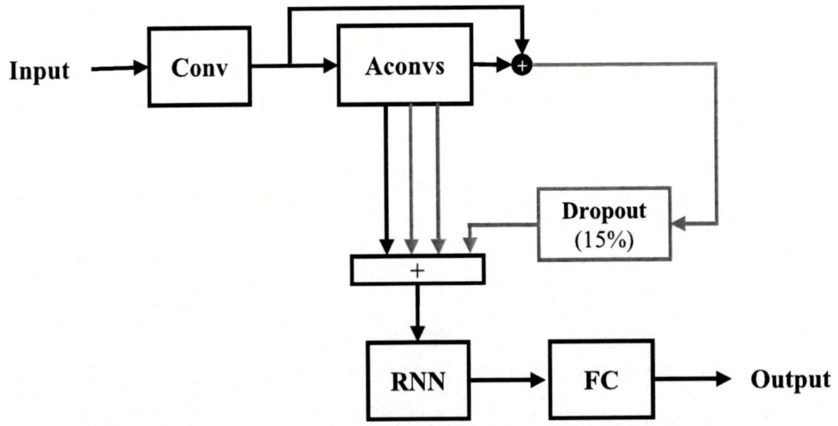

Fig. 7.2 Proposed feedback ANC controller

ANC system and acoustic paths. These extracted spatial features from each layer are summed before passing to the RNN with Tanh activations followed by a two-layer fully connected network with ten and one number of neurons, respectively. Finally, the output is the predicted anti-noise in the next elapsed time. Furthermore, in each layer of the network, skip connections are used to prevent gradient vanishing problems and ensure feature reusability [27], and a 15 percent dropout is applied to the final layer output to avoid overfitting. The total number of parameters of the design model is 345, making it suitable for realizing low latency real-time ANC system. Moreover, all operations used for designing the network are causal and do not need future data; hence the model is appropriate for real-time processing. Since this is not an adaptive algorithm and just requires one-time training, the model will not face the instability convergence problem of adaptive methods, for example, FxLMS, in real-time processing.

7.5 Data and Training Strategy

The data for training and testing the deep learning model is downloaded from [28]. This data includes military vehicle noise (Leopard) traveling at 70 km/h speed, vehicle interior noise (Volvo) while driving at 140 km/h on an asphalt road in the rain, and airplane cockpit noise. The data is recorded by the ½″ B&K condenser microphone with 19.98 kHz sampling frequency and 16 bits resolutions. NTiXL2 sound level meter with a type-1 NTi microphone is used to analyze the airplane noise and upsample it to 44.1 kHz sampling frequency. The data extracted from airplane cockpit noise is named Flight noise. After normalizing the data between −1 and 1, 10,000 and 5000 samples from all noise types were randomly selected for training and testing the network, respectively. We have ensured that the test data is completely different from the training data to verify the robustness and generalization of the model. The ANC algorithm is trained with a mean square error (MSE) loss function for 300 epochs, and the Adam optimizer [29] is utilized to update the network's trainable parameters in each iteration through backpropagation.

7.6 Results

In this section, the output results of the proposed FB ANC framework are presented in both time, frequency, and time-frequency domains to clearly visualize the effectiveness of the algorithm. Furthermore, the FB FxNLMS method [5], the most widely used ANC algorithm, has been utilized to compare the results. Choosing the suitable step size for the standard FxLMS algorithm is a bit hard, and the result of the algorithm is so sensitive to the value of step size. As a result, FxNLMS is used in this paper since it almost tackled this problem by a time-varying step size. For a fair comparison, the controller filter length is set a 20 same as the input length signal to the proposed ANC algorithm. The estimated FIR filter for the secondary path has 1500 samples equal to the length of simulated primary and secondary RIRs.

Figure 7.3 describes the noise mitigation results for Leopard and Flight noise types. Furthermore, Table 7.1 shows the noise attenuation in decibel (dB) obtained by the proposed and comparison algorithms calculated by Eq. (7.5), where E_N and

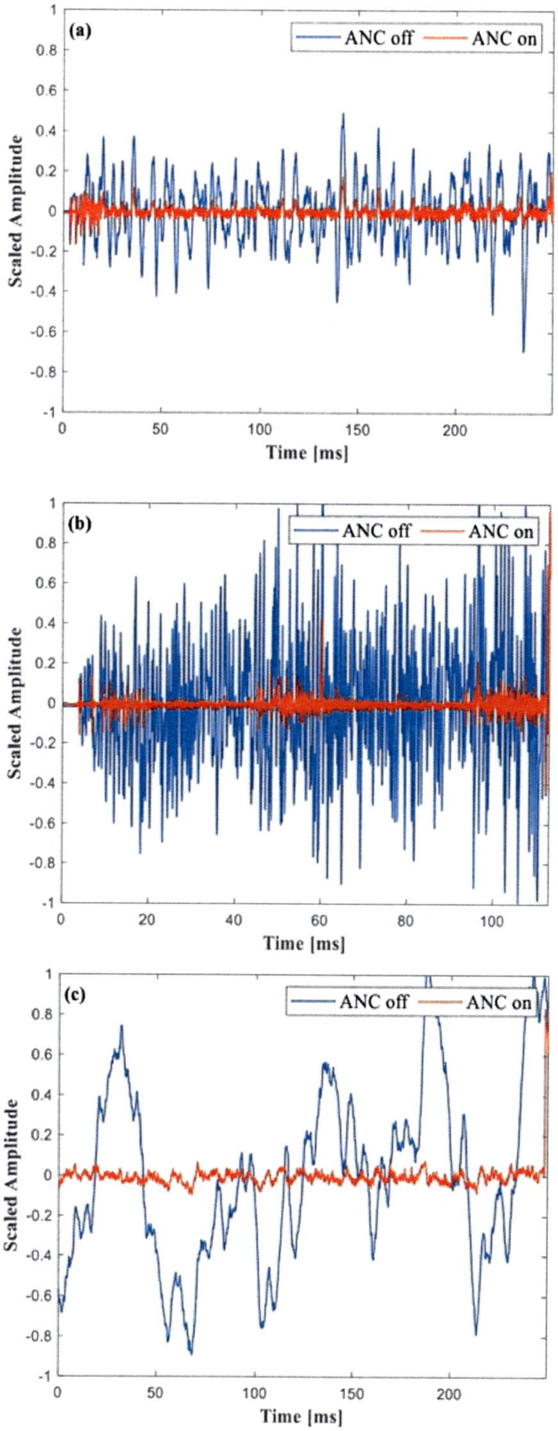

Fig. 7.3 Noise attenuation performance of the proposed FB ANC framework for (**a**) Leopard (**b**) Flight (**c**) Volvo

E_R are the energy of the primary and residual noise, respectively. In Fig. 7.3, one can compare the amplitude of attenuated and primary noise. It is clear that the majority of undesired original noise is eliminated by the proposed algorithm.

$$\text{Noise attenuation } (dB) = 10.\log_{10}.\frac{E_P}{E_R} \tag{7.5}$$

Regarding Table 7.1, as mentioned in the introduction section, the noise cancellation of FB FxNLMS is significantly inferior to the FB deep learning method because traditional adaptive ANC methods (FxNLMS being among the best) use

Table 7.1 Noise attenuation levels (dB) achieved by proposed ANC and FxNMLS with considering primary and secondary paths nonlinearities

Data/Method	Proposed ANC system	FxNLMS
Leopard	15.56	−0.06
Volvo	17.08	1.45
Flight	15.42	0.04
Average	**16.02**	**0.48**

linear filters for the controller and secondary path estimation, so they cannot deal with unavoidable inherent nonlinearities in noise, primary and secondary paths.

7.7 Conclusion

In this chapter, we have developed a new deep learning-based model for active noise cancellation in various noisy environments. 1D Atrous and casual convolution operations are carefully incorporated to extract effective features from estimated raw time-domain noise signals. Then multilevel features are processed by an RNN to predict anti-noise for upcoming noise samples by considering nonlinearities in primary and secondary paths. It is a lightweight model with just 345 parameters, suitable for real-time ANC applications. In addition, simulation results demonstrated that the proposed FB controller performed much better than the most well-known traditional ANC algorithm. In the next step, we will build an experimental setup to verify the algorithm's performance in real-word ANC applications.

Funding This research was partially supported by an NSERC Engage grant (Application No.: 533690 – 18), an NSERC Discovery Grant (RGPIN-2016–05923), a Research Manitoba Innovation Proof-of-Concept Grant (4914), and a CFI JELF grant (37394).

References

1. Schulz, A.J.: Urban environments and health A2. In: Nriagu, J.O. (ed.) Encyclopedia of Environmental Health, pp. 549–555. Elsevier (2011). Accessed 25 Aug 2022. [Online]. Available: http://www.sciencedirect.com/science/article/pii/B9780444522726000076
2. Gupta, S., Ghatak, C.: Environmental noise assessment and its effect on human health in an urban area. Int. J. Environ. Sci. **1**(7), 1954–1964 (2011)
3. Marburg, S.: Developments in structural-acoustic optimization for passive noise control. Arch. Comput. Meth. Eng. **9**(4) (2002). https://doi.org/10.1007/BF03041465
4. Lam, B., Gan, W.S., Shi, D.Y., Nishimura, M., Elliott, S.: Ten questions concerning active noise control in the built environment. Build. Environ. **200** (2021). https://doi.org/10.1016/j.buildenv.2021.107928
5. Kuo, S.M., Morgan, D.R.: Active Noise Control Systems Algorithms and DSP Implementations. A Wiley-Interscience Publication, New York/Chichester/Brisbane/Toronto/Singapore (1996)
6. Kestell, C.D., Australia, S.: Active control of sound in a small single engine aircraft cabin with virtual error sensors. In: Mechanical Engineering on (2000). Accessed 25 Aug 2022. [Online]. Available: https://hekyll.services.adelaide.edu.au/dspace/handle/2440/37700
7. Samarasinghe, P.N., Zhang, W., Abhayapala, T.D.: Recent advances in active noise control inside automobile cabins: toward quieter cars. IEEE Signal Process. Mag. **33**(6) (2016). https://doi.org/10.1109/MSP.2016.2601942
8. Dixit, S., Nagaria, D.: LMS adaptive filters for noise cancellation: a review. Int. J. Electr. Comput. Eng. **7**(5), 2520–2529 (2017). https://doi.org/10.11591/ijece.v7i5
9. Tobias, O.J., Seara, R.: Leaky-FXLMS algorithm: stochastic analysis for Gaussian data and secondary path modeling error. IEEE Trans. Speech Audio Proc. **13**(6) (2005). https://doi.org/10.1109/TSA.2005.852018
10. Ghasemi, S., Kamil, R., Marhaban, M.H.: Nonlinear Thf-Fxlms algorithm for active noise control with loudspeaker nonlinearity. Asian J Control. **18**(2) (2016). https://doi.org/10.1002/asjc.1140
11. Chen, C.K., Chiueh, T.D.: Multilayer perceptron neural networks for active noise cancellation. Proc. IEEE Int. Symp. Circuits Syst. **3** (1996). https://doi.org/10.1109/iscas.1996.541648
12. Na, K., Chae, S.I.: Single-sensor active noise cancellation using recurrent neural network predictors. In: *IEEE International Conference on Neural Networks – Conference Proceedings*, vol. 4, (1997). https://doi.org/10.1109/ICNN.1997.614239
13. Park, S., Patterson, E., Baum, C.: Long short-term memory and convolutional neural networks for active noise control. In: 2019 5th International Conference on Frontiers of Signal Processing (ICFSP) (2019). https://doi.org/10.1109/ICFSP48124.2019.8938042
14. Choi, W., Cha, Y.J.: SDDNet: real-time crack segmentation. IEEE Trans. Ind. Electron. **67**(9) (2020). https://doi.org/10.1109/TIE.2019.2945265
15. Kang, D.H., Cha, Y.J.: Efficient attention-based deep encoder and decoder for automatic crack segmentation. Struct. Health Monit. **21**, 1–16 (2021). https://doi.org/10.1177/14759217211053776

16. Cha, Y.J., Choi, W.: Deep learning-based crack damage detection using convolutional neural networks. Comput. Aided Civil Inf. Eng. **32**(5), 361–378 (2017, Wiley Online Library). https://doi.org/10.1111/mice.12263
17. Cha, Y., Choi, W., Suh, G., Mahmoudkhani, S., Büyüköztürk, O.: Autonomous structural visual inspection using region-based deep learning for detecting multiple damage types. Comput. Aided Civil Inf. Eng. **33**(9), 1–17, Wiley Online Library (2018). https://doi.org/10.1111/mice.12334
18. Kang, D., Cha, Y.J.: Autonomous UAVs for structural health monitoring using deep learning and an ultrasonic Beacon system with geo-tagging. Comput. Aided Civ. Inf. Eng. **33**(10) (2018). https://doi.org/10.1111/mice.12375
19. Beckman, G.H., Polyzois, D., Cha, Y.J.: Deep learning-based automatic volumetric damage quantification using depth camera. Autom. Constr. **99** (2019). https://doi.org/10.1016/j.autcon.2018.12.006
20. Wang, Z., Cha, Y.J.: Unsupervised deep learning approach using a deep auto-encoder with a one-class support vector machine to detect damage. Struct. Health Monit. **20**(1) (2021). https://doi.org/10.1177/1475921720934051
21. Maasoum, S.M.H., Mostafavi, A., Sadighi, A.: An autoencoder-based algorithm for fault detection of rotating machines, suitable for online learning and standalone applications. IEEE (2020). https://doi.org/10.1109/ICSPIS51611.2020.9349574
22. Zhang, H., Wang, D.L.: Deep ANC: a deep learning approach to active noise control. Neural Netw. **141** (2021). https://doi.org/10.1016/j.neunet.2021.03.037
23. Mostafavi, A., Sadighi, A.: A Novel Online Machine Learning Approach for Real-Time Condition Monitoring of Rotating Machines. In: 2021 9th RSI International Conference on Robotics and Mechatronics (ICRoM). ieeexplore.ieee.org, Accessed 26 Aug 2022. [Online]. Available: https://ieeexplore.ieee.org/abstract/document/9663495/ (2021)
24. Habets, E.: Room Impulse Response Generator. International Audio Laboratories Erlanggen (2010)
25. Allen, J.B., Berkley, D.A.: Image method for efficiently simulating small-room acoustics. J. Acoust. Soc. Am. **65**(4) (1979). https://doi.org/10.1121/1.382599
26. Elman, J.L.: Finding structure in time. Cogn. Sci. **14**(2) (1990). https://doi.org/10.1016/0364-0213(90)90002-E
27. He, K., Zhang, X., Ren, S., Sun, J.: Deep residual learning for image recognition. In: *Proceedings of the IEEE Computer Society Conference on Computer Vision and Pattern Recognition*, vol. 2016, (2016). https://doi.org/10.1109/CVPR.2016.90
28. Signal processing Information Base (SPIB). Accessed 20 June 2022. [Online]. Available: http://spib.linse.ufsc.br/
29. Kingma, D.P., Ba, J.L.: Adam: A method for stochastic optimization. arXiv preprint arXiv:1412.6980 (2014). https://arxiv.org/abs/1412.6980

Chapter 8
Govan-Partick Pedestrian Bridge: Piezoelectric Energy Harvesting from Footfall-Induced Vibrations

Venkatsubramaniam Shashank, Falcone Gioia, and Cammarano Andrea

Abstract The Govan-Partick pedestrian bridge is an under-construction footbridge that aims to reconnect the less developed Govan region of Glasgow to the neighboring Partick region, which contains cultural, economic, and educational landmarks, such as the University of Glasgow. This chapter aims to simulate the total energy that can be harvested via footfall-induced vibrations by people and cycles travelling on the bridge and display the data to the public to encourage active travel. This is done in accordance with the University's GALLANT (Glasgow as a Living Lab Accelerating Novel Transformation) Project, which aims to increase active travel, such as walking and cycling, within Glasgow and aim for low carbon energy solutions.

Piezoelectric energy harvesting is a growing field that utilizes mechanical vibrations and stresses to derive electric energy. Using piezo patches installed on cantilever beams, the vibrations induced from footfall are harvested to generate an electric potential, which can then be stored or used to charge sensors used to monitor the activity on the bridge. The data will be displayed in real time to promote active travelling.

By modelling the bridge in a Finite Element Analysis Software, this work aims to derive the modal measurements of the bridge and import them into a simulation program to obtain the potential energy that can be generated under different parametric conditions. This information is core to correctly dimension the sensor network and the visualization device. Preliminary energy estimations will be driven by data obtained from observation of the footfall on already existing footbridges across the river Clyde.

Keywords Piezoelectric energy harvesters · Vibrations · Finite element analysis · Footbridge · Active travel · GALLANT

8.1 Introduction

The impact of the COVID 19 pandemic on the energy sector led to a decrease of 5.1% in global Carbon emissions, as reported by the International Energy Agency's Global Energy Emissions report for the year 2020 [19]. A rapid global economic recovery, boosted by a rapid roll out of vaccines, has increased the energy demand, leading to a rise in the consumption of fossil fuels, with CO_2 emissions surpassing prepandemic levels. An increased usage in vehicular transport leads to further concerns of carbon emissions, with vehicles historically contributing nearly 30% of the total carbon emissions. The IEA projects transport related emissions to add nearly 600 Megatons to worldwide CO_2 emissions [19]. This leads to serious health problems due to declining air quality and carbon impacts from travel by vehicles.

This concern is shared by the Glasgow City Council, as seen in their framework for Glasgow's Transport Strategy [20]. As mentioned by the City Convenor for Sustainability & Carbon Reduction, sustainability is one of the main goals of the Strategy framework, as they hope to help meet the City Council's goal of making Glasgow Net Zero Carbon by 2030.

V. Shashank (✉)
James Watt School of Engineering, University of Glasgow, Glasgow, UK
e-mail: shash98@gmail.com

F. Gioia
Rankine Chair of Energy Engineering (Systems Power & Energy), University of Glasgow, Glasgow, UK
e-mail: gioia.falcone@gla.ac.uk

C. Andrea
James Watt School of Engineering & Marine Engineering, University of Glasgow, Glasgow, UK
e-mail: andrea.cammarano@gla.ac.uk

C. Walber et al. (eds.), *Sensors & Instrumentation and Aircraft/Aerospace Testing Techniques, Volume 8*, Conference Proceedings of the Society for Experimental Mechanics Series, https://doi.org/10.1007/978-3-031-34938-6_8

Policy 1 of the framework specifies the need to reduce the total number of short trips and to deliver the concept of 20-min neighborhood via economic policies. The policy hopes to result in an increased trend in active travel, such as cycling and walking. This would further tackle the secondary issue of reducing car usage for short distances from its current 27% rate [20]. This project hopes to monitor the total energy harvested due to the footfalls of a person walking across the selected span and showcase this energy using a monitor screen. This study hopes that this would lead to a positive reinforcement cycle regarding usage of active travel methods and also test the feasibility of using piezoelectric harvesters on civil infrastructures.

The GALLANT (Glasgow as a Living Lab Accelerating Novel Transformation) project is a joint collaboration between the University of Glasgow and the Glasgow City Council that aims to utilize Glasgow as a living lab project to test and estimate the success of various sustainable solutions and experiments [16]. The main aim of this £10.2 million project is to strike a balance between the various aspects that influence public health, well-being, and the economy. Five Work Packages (WP) have been created under this project to achieve the goal of making Glasgow carbon neutral by 2030. WP4 of GALLANT aims to promote active travel such as cycling and walking and reduce overall carbon emissions, while WP 5 aims to create low carbon energy solutions that utilize the citizens as active contributors [16]. Considering the aims of this study, it becomes closely associated with the abovementioned work packages.

The main principle behind vibration energy harvesting is in using mechanical vibrations, stresses, and strains to generate electric potential. These mechanical vibrations can be intended vibrations, such as via using pressure plates or by using waste vibrations from contact forces from vehicles passing through or from footfall motions. There are primarily three methods to achieve this: by using a magnetic field (electromagnetic energy harvesting), by using an external electric field (electrostatic energy harvesting), or by using a piezoelectric material (piezoelectric energy harvesting) [4]. This study focuses on the third type, which is piezoelectric energy harvester (PEH).

Piezoelectric material has the property of producing an electric potential on being strained and vice versa. Piezoelectric energy harvesters utilize the direct piezoelectric effect, wherein an electric potential is generated on applied strain [18]. Piezoelectric energy harvesting has an inherent advantage at low frequency vibrations as it is able to capture energy from low frequency vibrations and convert them to electricity [5]. However, maximum energy is captured when the frequency of the forced vibrations is in the range of the resonant frequency of the energy harvester [3]. Research conducted by Ye Zhang studied the feasibility of using piezoelectric energy harvesters in bridge structures to harvest the vibrations induced from vehicular motion. The study concluded that it was feasible for a PEH system to power a wireless sensor node and further studied the various parameters the affect the potential energy harvested [1].

The main drawback of using a piezoelectric energy harvester in a bridge system is the mismatch between the resonant frequency of the piezo crystals and the natural frequency of the bridge system itself [1]. This disparity can be reduced by reducing the stiffness of the PEH, by attaching the patch on a cantilever substructure. This unimorph configuration leads to a reduced resonant frequency, which can further be reduced by attaching a proof mass at the tip of the cantilever. Studies have shown that the eigen-frequency of the first and second node can be reduced by 58.01% and 21.78%, respectively, by attaching a proof mass [8].

8.1.1 Govan Partick Bridge

The Glasgow City Development Plan (GCDP) aims to create spatial conditions to improve economic conditions and reduce connectivity issues on either side of the river Clyde in Glasgow [14]. The Govan-Partick region is marked as one of the six central regions, due to the region being characterized by an uneven pattern of educational, cultural and economic development. The Strategic Development Framework adopted by the GCDP in 2017 regards the under-construction Govan – Partick bridge as an important link to increase development and economic opportunities on either side of the river Clyde [14].

Commissioned to Farran's (Construction) Ltd., the footbridge connects the Water Row in the less developed Govan region to the Pointhouse Quay in the more affluent Partick region. The City Council estimates the construction of the footbridge to have an increased impact on the uptick of job opportunities, with a projection of a 23% increase in job opportunities that are within a 20-min walk of the Govan Cross [13]. One of the longest opening footbridges in Europe, the 115 m (about 377.3 ft) long bridge is divided into three spans and is cable stayed.

The 68 m central span is the focus of this study's research and hosts a 28.5 m tower where the cables supporting the bridge terminate. Both the central span and the north span (towards Partick) have a steel deck that is supported by two inlined steel box girders. The south span has a deck that is reinforced concrete and is placed over two steel box girders [12]. This study assumes that the load bearing effect of the box girders are negligible, and the central span is supported by the steel cables.

8.2 Methodology

8.2.1 Bridge Model

The bridge deck was modelled in SOLIDWORKS and was imported to ABAQUS, where the bridge cables were modelled by kinematically coupling certain nodes of the bridge deck to fixed points in space above the bridge deck, based on the actual start and end points of the cables as given in the drawing plan of the bridge [15]. The nodes on the bridge deck were coupled such that their translational degrees of freedom with respect to the motion of the fixed points were restricted, that is the selected nodes were only allowed to rotate [17]. The mesh selected for this study was a linear wedge mesh (C3D6), as it allowed us to use the least computing resources as possible, but the accuracy of the study was not impacted due to the simplicity of the structure.

8.2.2 Bridge-Person Interaction

The bridge can be likened to a damped forced mass-spring system with Multiple Degrees of Freedoms (MDOFs) [1]. The forcing caused to the footfalls of the person walking is considered sinusoidal in nature. The equations representing the bridge are given as:

$$[M]\{\ddot{x}\} + [C]\{\dot{x}\} + [K]\{x\} = \{F(t)\} \tag{8.1}$$

where $[M]$, $[C]$, and $[K]$ represent the mass, damping, and stiffness matrices while $\{F(t)\}$ is the force vector caused by the subject's footfalls. This force vector is given by Eq. (8.2) as [6]:

$$F(t) = P\alpha_i{}^{*} \cos\left(2\pi \mathrm{i} f_{\mathrm{step}} t\right) \tag{8.2}$$

where P is the weight of the subject, f_{step} gives the step frequency, while α_i is the dynamic coefficient for the ith force component. The value for the dynamic coefficient and the force component is dependent on the forcing frequency, and the values are represented in Table 8.1 [7].

For this particular study, only one test subject was considered, and the parameters of the test subject are given in Table 8.2.

8.2.3 Piezoelectric Energy Harvester

A unimorph cantilever energy harvester refers to a piezoelectric energy harvester wherein a single piezoelectric patch is attached to a substructure to reduce the stiffness of the piezoelectric material and, thus, attune its reduced resonance frequency to the frequency of the forcing vibrations [1]. The proof mass is used to further reduce the stiffness and, thus, the natural

Table 8.1 Dynamic Coefficients & Forcing Frequencies

Forcing frequency (Hz)	Harmonic multiple	Dynamic coefficient
1.6–2.2	1	0.5
3.2–4.4	2	0.2
4.8–6.6	3	0.1
6.4–8.8	4	0.05

Table 8.2 Walking Parameters of Test Subject

Parameter	Value
Weight	62.9 kg
Velocity	1.3 m/s
Stride length	0.74 m
Step frequency	1.74 Hz

frequency of the energy harvester. This is important for a piezoelectric energy harvester as maximum energy can be harvested if the incident vibrations are in the same frequency range as that of the energy harvester. Studies have shown that addition of a proof mass can significantly reduce the resonant frequency of an energy harvester [8].

On addition of a proof mass, the resonant frequency of a unimorph cantilever for the *n*th node can be written as [3]:

$$\text{fn} = \left(\upsilon_n{}^2/2\pi\right)^* \text{sqrt}\ (K/m_e + \text{PM}) \tag{8.3}$$

where υ_n is the eigen value of the corresponding node, K is the stiffness of the unimorph, and m_e and PM are the effective mass of the cantilever and the proof mass, respectively [10].

The stiffness of the unimorph can be calculated as a function of the thicknesses (t_p & t_s), Young's Moduli (E_p & E_s), and the dimensions of the unimorph (l & w) as [3]:

$$K = 3D_p w/l^3 \tag{8.4}$$

$$D_p = \left[E_p{}^2 t_p{}^4 + E_s{}^2 t_s{}^4 + 2E_p t_p E_s t_s{}^* \left(2t_p{}^2 + 2t_s{}^2 + 3t_p t_s\right)\right] /12^* \left(E_p\, t_p + E_s\, t_s\right) \tag{8.5}$$

If the resonant frequency of the unimorph matches with the frequency range of the forcing vibrations, then the energy harvester can be simplified as a damped mass-spring system with a Single Degree of Freedom (SDOF) [2]. In such a case, the energy harvester can be represented by an electromechanical coupled equation [11], given by Eqs. (8.6a) and (8.6b)

$$m\ddot{x} + c\dot{x} + Kx – k_{\text{me}}\ V = \{f(t)\} \tag{8.6a}$$

$$q – k_{\text{me}}\ x – \left(1/c_p\right)\ V = 0 \tag{8.6b}$$

where k_{me} represents the electromechanical coupling [9], q represents the charge, c_p is the inherent capacitance of the system, and V is the voltage.

If we were to consider that the system is connected to a purely resistive load, then the charge and voltage can be written as:

$$\text{R}\dot{\text{q}} + V = 0 \tag{8.7}$$

By substituting Eq. (8.7) in Eqs. (8.6a) and (8.6b), the coupled equations can be rewritten as:

$$m\ddot{x} + c\dot{x} + \left(K + k_{\text{me}}{}^2/c_p\right)\ x – \left(k_{\text{me}}/c_p\right)\ q = \{f(t)\} \tag{8.8a}$$

$$R\ \dot{q} + \left(1/c_p\right)\ q – \left(k_{\text{me}}/c_p\right)\ x = 0 \tag{8.8b}$$

On solving Eqs. 8.8a and 8.8b, we can determine the charge collected on the cantilever surfaces as a function of time.

8.3 Analysis & Results

8.3.1 Govan-Partick Bridge

The modelled bridge deck with the cables is given in Fig. 8.1. The reference points RP1 to RP10 represent the nodes on the bridge deck, while the reference points RP11 to RP15 represent the fixed points in space. The two ends of the bridge deck are constrained in place using a Boundary Condition to represent the bridge piles.

On running a frequency analysis for the first 10 eigen modes, the corresponding eigen frequencies are given in Table 8.3 and the shape of the first three eigen-modes are given in Figs. 8.2, 8.3, and 8.4.

From the three eigen shapes, we can conclude that the optimal position for placing the energy harvester is at the midpoint of the span and the operating frequency for the piezoelectric energy harvester should be in the frequency range of 0.8–

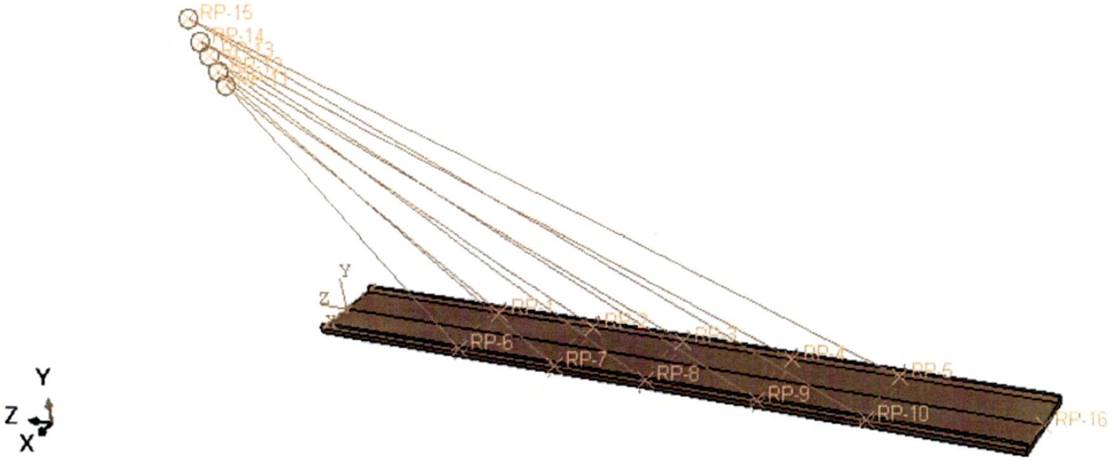

Fig. 8.1 Modelled bridge deck with cables

Table 8.3 Frequencies of first 10 Eigen Modes

Mode number	Eigen frequency
1	0.7988
2	2.1960
3	3.3064
4	4.2892
5	6.7710
6	7.0603
7	8.7084
8	10.491
9	10.533
10	14.553

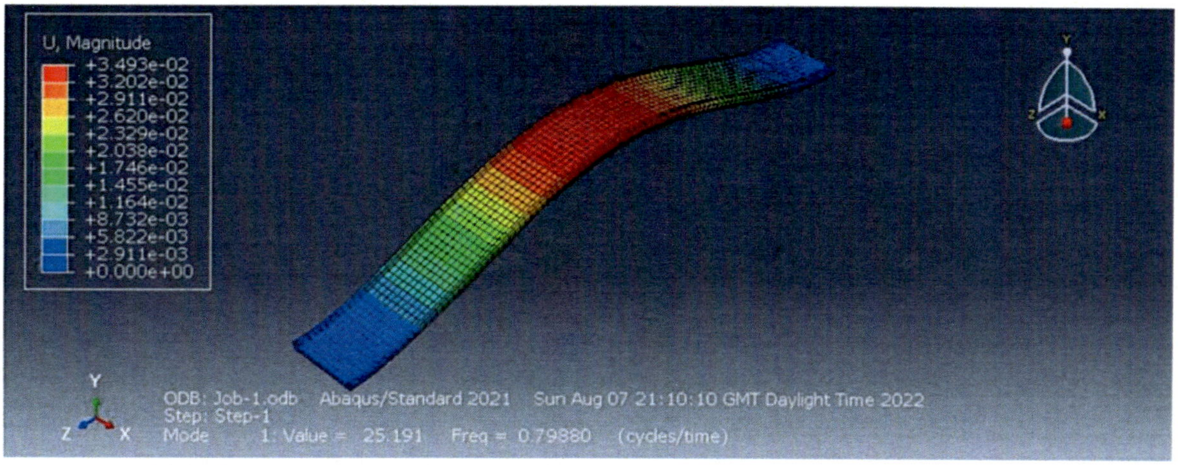

Fig. 8.2 Eigenshape of the first eigen-mode

2.2 Hz. This is due to the first eigen-mode being the most optimal mode for the vibrations to be captured, as seen from the eigen-shapes.

On substituting the value of the optimal frequency in Eq. (8.3), the parameters of the energy harvester can be obtained, as provided in Table 8.4.

By obtaining the value of the forcing frequency from Table 8.2, we can obtain the values of the dynamic coefficient and the harmonic multiple from Table 8.1, which are as follows:

$i = 1; \alpha_1 = 0.5.$

By substituting these values in Eq. (8.2), we can rewrite the time-based force as:

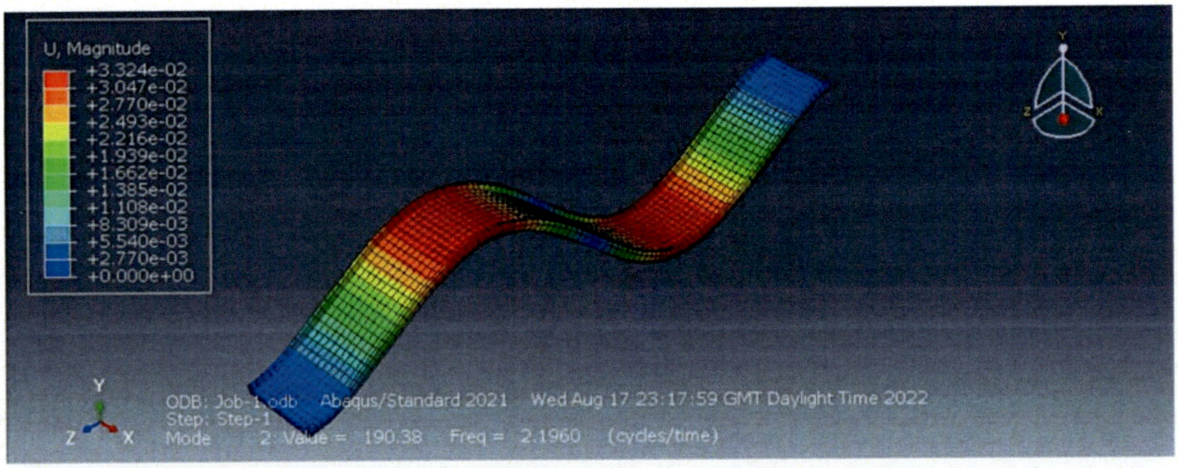

Fig. 8.3 Eigenshape of the second eigen-mode

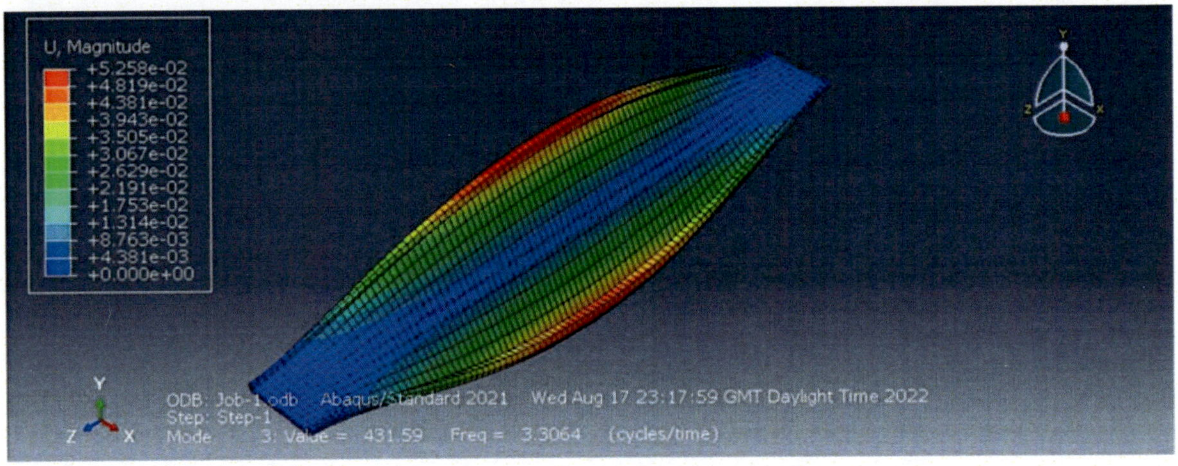

Fig. 8.4 Shape of the third eigen-mode

Table 8.4 Dimensions of Unimorph Cantilever Beam Model

Parameter	Unit	Value
Length of the unimorph	l	70 mm
Width of the unimorph	w	30 mm
Thickness of the piezoelectric patch	t_p	1 μm
Young's modulus of the piezoelectric patch	E_p	75 GPa
Thickness of the substructure	t_s	1.2 μm
Young's modulus of the substructure	E_s	69.8 GPa
Density of the piezoelectric patch	ρ_p	7500 kg/m^3
Density of substructure	ρ_s	2750 kg/m^3
Proof mass	PM	0.6 kg

$$F\ (t) = 62.8^{*}9.8^{*}0.5^{*}\cos\left(2\pi^{*}1^{*}1.74^{*}t\right) = 303.8\cos\left(10.94t\right) \tag{8.9}$$

By applying the amplitude and frequency of this force on the bridge in ABAQUS, we can determine the nodal displacement of the PEH as a function of time, as shown in Fig. 8.5:

The forcing caused due to the nodal vibrations is determined by double differentiating the displacement data with respect to time and is inputted into the coupled ODEs given in Eqs. 8.8a and 8.8b. On solving the ODEs with the given input, the charge collected with respect to the time spent by the subject in traversing the span, as seen in Fig. 8.6.

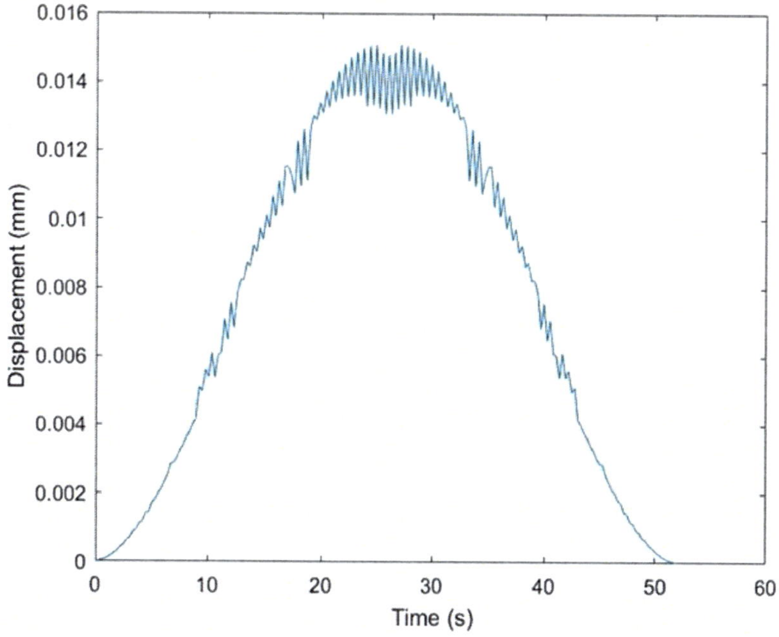

Fig. 8.5 Displacement vs time graph

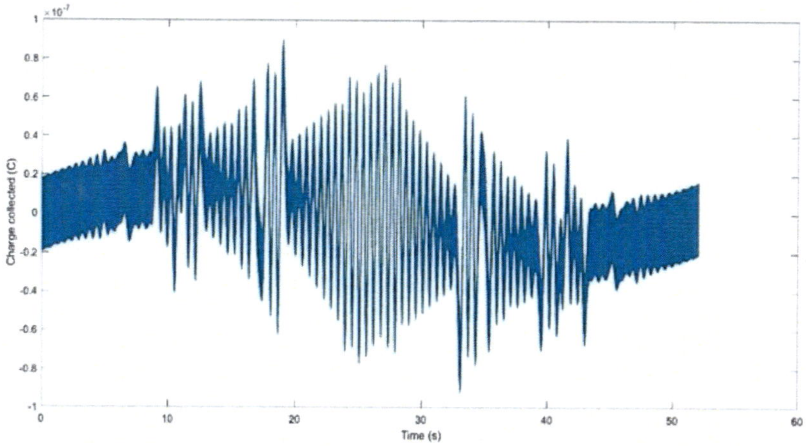

Fig. 8.6 Charge collected vs time graph

If we treat the unimorph as a dielectric material, the voltage generated can be calculated using the inherent capacitance of the piezoelectric material and the total power generated can be written as:

$$P_{\text{out}} = V^2/R_{\text{total}} = q^2/\left(c_p{}^* R_{\text{total}}\right)^2 \tag{8.10}$$

The output power, when plotted against time, is as given in Fig. 8.7.

The total power output can be calculated by integrating the above graph with respect to time and the total output power comes as:

$P_{\text{out}} = 6.604$ mW in 52 seconds

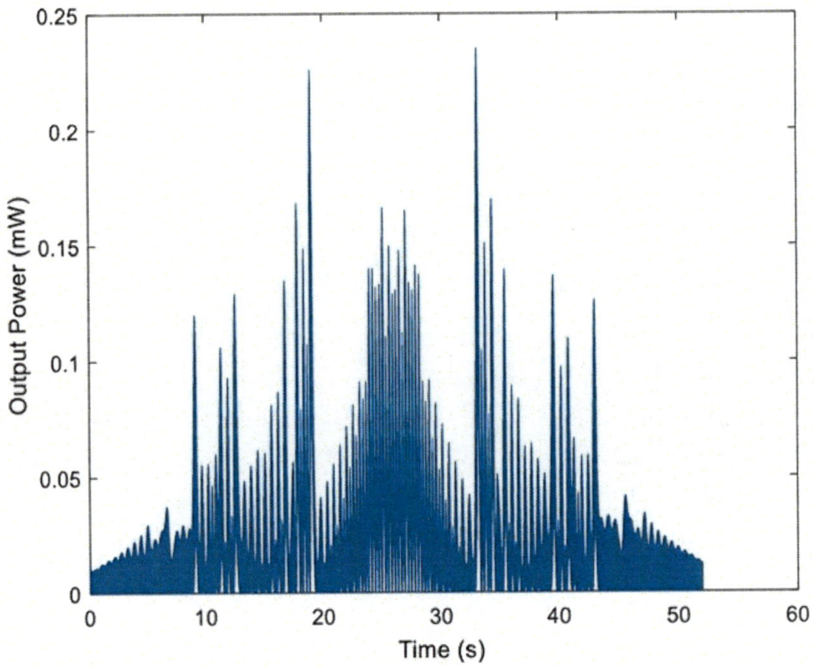

Fig. 8.7 Output power vs time graph

8.4 Conclusion

This chapter tried to simulate the total power that could be harvested by the vibrations induced by the footfall of a single person walking on the under-construction Govan-Partick bridge. To accomplish this, the bridge deck was modelled in SOLIDOWRKS with the cables further being modelled in ABAQUS using the constraint "Kinematic Coupling." The eigen-frequencies of the first 10 eigen-modes were determined, and the operating frequency was determined using the mode shape. The optimal position of the piezoelectric energy harvester was also determined using the mode shapes.

The time-dependent force due to the test subject's walking was calculated and inputted in ABAQUS to obtain the nodal displacement, which was used to determine the forcing vibration, and a unimorph cantilever beam model was developed to represent the energy harvester. Based on the operating frequency obtained from ABAQUS, the dimensions of the cantilever beam were decided and so was the material of the piezoelectric patch and the substructure. The nodal displacements were used to obtain the force vector to be imposed on the unimorph beam model. The coupled ODEs were solved in MATLAB using the ode45 solver, and the state space models of the ODEs and the instantaneous charge collected on the piezoelectric surfaces were obtained. This data was used to determine the voltage, and correspondingly, the maximum and the total output power was obtained.

The results obtained were in the expected range of energy harvested using piezoelectric material due to the low frequency vibrations in civil structures. This chapter can, thus, conclude that the usage of a piezoelectric energy harvester for a civil structure would not be a feasible project. The project recommends using an electromagnetic energy harvester instead, which would be better suited for capturing low frequency vibrations present due to the large inherent damping of civil structures. The chapter would also recommend exploring the feasibility of using multiple energy harvesters to better capture the vibrations.

The chapter also notes the challenge of using a large proof mass to reduce the frequency of the cantilever and recommends alternative solutions to reach a compromise between having a low operating frequency and a smaller proof mass compared to the dimensions of the energy harvester.

References

1. Zhang, Y.: Piezoelectric Based Energy Harvesting on Low Frequency Vibrations of Civil Infrastructures. LSU Doctoral Dissertations. 1342 (2004)
2. Lu, F., Lee, H.P., Lim, S.P.: Modelling and analysis of micro piezoelectric power generators for micro electromechanical systems applications. Smart Mater. Struct. (2003). https://doi.org/10.1088/0964-1726/13/1/007
3. Shen, D., et al.: The design fabrication & evaluation of a MEMS PZT cantilever with an integrated Si proof mass for vibration energy harvesting. J. Micromech. Micro Eng. **18**, 055017 (2008)
4. Wei, C., Jing, X.: A comprehensive review on vibration energy harvesting: modelling and realization. Renew. Sust. Energ. Rev. **74**, 1–18 (2007)
5. Na, L., Yuhao, W., Huanqing, H., Tongshou, L.: A review on vibration energy harvesting. E3S Web Conf. **245**, 01041 (2021)
6. Aw, J.H., Parvez, M.: Effect of footfall induced vibrations on footbridges. Appl. Mech. Mater. **802**, 136–141 (2015)
7. Murray, T.M., Allen, D.E., Ungar, E. E.: Floor Vibrations Due to Human Activity – DG-11 (10M797) (2001)
8. Salunke, S.V., Roy, S., Jagtap, K.R.: Modeling of piezoelectric energy harvester and comparative performance study of the proof mass for Eigen frequency. Mater. Today Proc. **5**, 4309–4316 (2018)
9. Tung, V.T., Tinh, N.T., Yen, N.H., Tuan, D.A.: Evaluation of electromechanical coupling factor for piezoelectric materials using finite element modeling. Int. J. Mater. Chem. **3**(3), 59–63. ISSN: 2166-5346 (2013)
10. Bucher, P.: Eigenvalue Computation for a Cantilever Beam. https://kratosmultiphysics.github.io/Examples/structural_mechanics/validation/beam_eigenvalue_analysis/
11. Erturk, A., Inman, D.J.: Piezoelectric Energy Harvesting, 1st edn. Wiley, Chichester (2011)
12. KGAL Consulting Engineers: FLUID, Autumn (2020)
13. Glasgow City region: Final Design for Govan – Partick Pedestrian Bridge Now Complete. https://glasgowcityregion.co.uk/final-design-for-govan-partick-pedestrian-bridge-now-complete/
14. Glasgow City Council: Govan Partick Strategic Development Framework, February (2020)
15. Glasgow City Council: Glasgow City Council Govan Partick opening bridge Scheme. General Arrangement – Bridge Closed. https://www.glasgow.gov.uk/CHttpHandler.ashx?id=48856&p=0 (2020)
16. University of Glasgow: Gallant: Glasgow as a Living Lab Accelerating Novel Transformation. https://www.gla.ac.uk/schools/education/research/fromlocaltoglobalresearch/researchprojects/gallant/
17. Smith, M.: ABAQUS/Standard User's Manual, Version 6.9. Dassault Systèmes Simulia Corp, Providence, RI (2009)
18. Priya, S., Inman, D.J.: Chapter 1: Energy Harvesting Technology. Springer (2009)
19. International Energy Agency: Global Energy Review: CO2 Emissions in 2021 (2022)
20. Glasgow City Council: Glasgow Transport Strategy Framework (2022)

Chapter 9
OASIS: Open Acquisition System for IEPE Sensors: For Academic Research and Teaching Purposes

Oliver M. Zobel, Johannes Maierhofer, and Daniel J. Rixen

Abstract Expensive measurement equipment often inhibits students from gaining practical experience with vibration measurements. The commercially available, proprietary hardware and software additionally does not grant insight into the used algorithms or allow modification and expansion. Both issues are overcome with the use of community developed open-source designs. This chapter discusses the requirements for vibration measurements in an academic context, like sampling frequency or resolution and especially synchronicity. These requirements are derived and incorporated into the design of an open-source data acquisition board for IEPE sensors (OASIS). The design is focused on the use of commonly available parts and a broadly community supported micro-controller (ESP32 family), with the costs bounded below 100 euro.

The built acquisition system offers four channels for IEPE signals, which can be sampled with up to 20 kHz and 16-bit resolution continuously. Additionally, wireless synchronization of multiple data acquisition boards is provided. This allows for use cases on systems with partly rotating structures, e.g., wind turbines, where synchronous measurements of stationary and rotating parts are necessary. The here-proposed approach does not rely on external clocks, like GPS or network services. An experimental validation shows that it is possible to synchronize two systems using this approach with a delay that is less than 100 μs.

Keywords Experimental dynamics · Data acquisition · Vibration analysis · Data synchronization · Measurement equipment

9.1 Introduction

Teaching of vibration measurement at universities is often limited to a theoretical introduction, missing the practical application in general. One factor for this might be the quite expensive measurement equipment, which makes large lab courses with equipment for all students impossible. Prices for small 4-channel data acquisition systems, which are the focus of this chapter, can reach into the high thousands of euros, not including the required, proprietary software.

This leads to the idea to design and build self-developed, cheap measurement hardware and software. Of course, this kind of equipment cannot match the accuracy, robustness, or reliability of commercially available products, but it possesses other advantages. Besides the considerably lower costs, it enables complete insight into the hardware and software design. This allows for adaptions, like additional hardware modules or new software features. Providing such hardware and software as open-source extends the concept even further, allowing a broad community to contribute. A solution for these demands was developed at the Chair of Applied Mechanics over the last 3 years. Figure 9.1 shows the latest prototype of the data acquisition system, which will be detailed in this chapter.

The goal of this chapter is to derive the general requirements of data acquisition systems for vibration measurements in an academic context. This includes basics, like sampling frequency or resolution, but the main focus is the required synchronicity between the acquired signals. To discuss synchronicity, a distinction needs to be made between the synchronous start of all measurements and sampling all signals with the same frequency. An illustration of the two effects is given in Fig. 9.2.

Recordings that start incorrectly, but with an exact sample interval (blue), lead to a constant phase shift in the reconstructed signal with respect to the original signal, resulting in altered phase relations with other observed signals. The start time offset

---

O. M. Zobel (✉) · J. Maierhofer · D. J. Rixen
Chair of Applied Mechanics, TUM School of Engineering and Design, Technical University of Munich, Garching, Germany
e-mail: oliver.zobel@tum.de; j.maierhofer@tum.de; rixen@tum.de

C. Walber et al. (eds.), *Sensors & Instrumentation and Aircraft/Aerospace Testing Techniques, Volume 8*, Conference Proceedings of the Society for Experimental Mechanics Series, https://doi.org/10.1007/978-3-031-34938-6_9

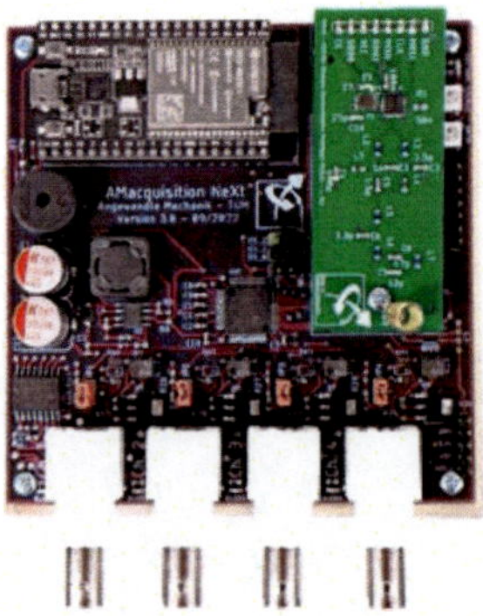

Fig. 9.1 Self-developed acquisition system

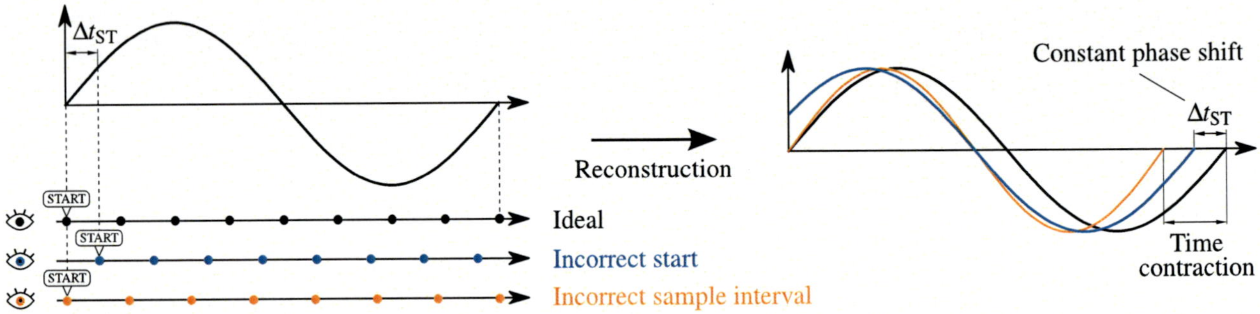

Fig. 9.2 Illustration of the two kinds of synchronicity errors and arising issues after signal reconstruction

Δt_{ST} should therefore be minimal. Measurements that start perfectly synchronous, but with an incorrect sample interval (orange), e.g., from different reference clocks, lead to a reconstructed signal that is either contracted or stretched in the time dimension. The cause of this error is described as drift per second τ_{drift}, the time added or subtracted from one true second of time. A positive drift means that the sample interval is too long, and the fixed number of samples, derived from sampling frequency and measurement duration, takes longer to acquire. All these considerations do not include sampling jitter, defined as the random deviation from the exact sampling interval [1].

Simultaneous sampling of multiple inputs within one Analog–Digital Converter (ADC) is ready to use for high-quality devices. But connecting multiple devices together needs careful attention to provide synchronicity. While synchronization of multiple systems via cable is commonly available, wireless systems often rely on external clocks like GPS or network services. Both methods use absolute timestamps added to the signal in order to synchronize the acquired data in post-processing. The shortcoming of GPS lies in the required reception of GPS signals, which might be troublesome indoors, while the implementation of time-sensitive networking is quite complex. Here, an approach for wireless synchronization of data acquisition systems is proposed that is self-contained and nearly without overhead. This allows for use cases on systems with partly rotating structures, e.g., wind turbines, where synchronous measurements of stationary and rotating parts are necessary.

From the derived requirements, an acquisition system is built that is focused on the use of commonly available parts and a broadly community supported micro-controller (*Espressif ESP32*). To achieve affordability, the costs are bounded below 100 euro. Additionally, the system is open-sourced to enable everyone to use, understand, and extend its functionality.

9.2 Performance Requirements for Data Acquisition

The main step of data acquisition is the conversion of the analog measurement signals to digital values. During this, the signals are discretized in time and amplitude (quantization) [2]. This raises two choosable parameters for an acquisition system: the (maximum) sample rate f_S and the quantization resolution n_{ADC}. While the latter solely depends on the specifications of the chosen ADC, the former depends on the whole system performance. With the focus being modal analysis, substructuring, or similar evaluations with the acquired data, the phase relation of the different signals as well as accurate frequency reconstruction is important. As described in the introduction, this introduces two additional requirements:

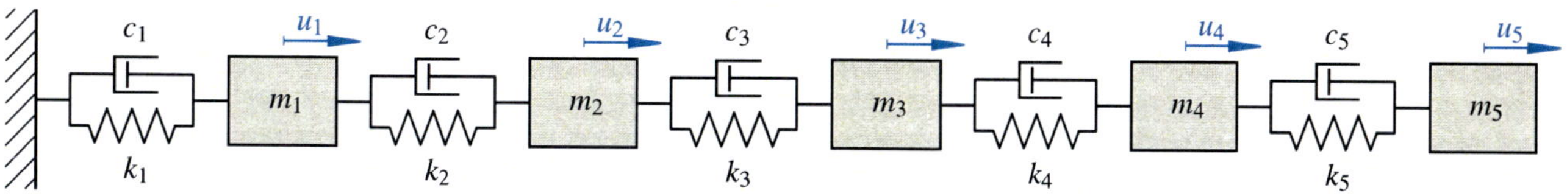

Fig. 9.3 Lumped mass oscillator model for the investigation of synchronicity requirements, with five degrees of freedom u_i

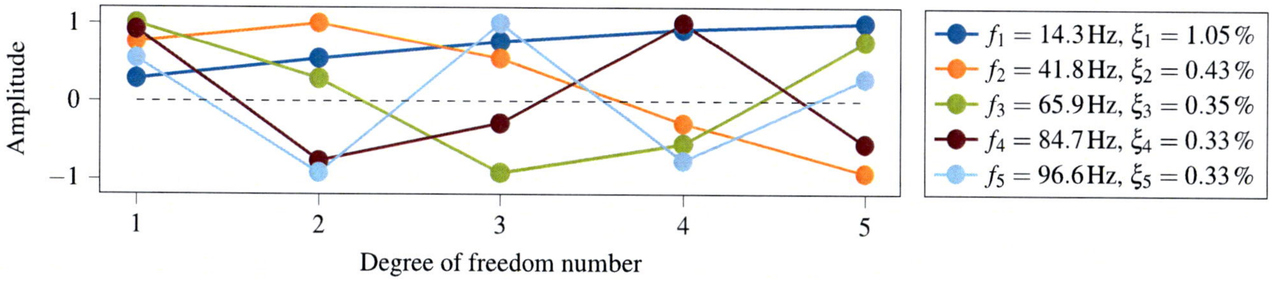

Fig. 9.4 Original mode shapes of the five mass oscillators with eigenfrequencies f_i and modal damping ratio ξ_i

the maximally allowed start time offset Δt_{ST} and drift per second τ_{drift}. Before designing the hardware, these requirements are derived individually.

Sample rate For the validation of new methods, e.g., for experimental substructuring on aluminum structures as in [3], frequency response functions up to 3 kHz are of interest. Adding additional headroom to avoid the area around the cutoff frequency of any low-pass filters, 5 kHz bandwidth, and with the Nyquist–Shannon theorem a sampling frequency f_S of at least 10 kHz is desired. Since most cheap ADCs can handle sampling frequencies of 50 kHz and above, this requirement mostly applies to the software driving the ADC, rather than the hardware itself. For applications with impulse hammers, a higher sample rate might be of interest in order to fully capture the impulse, especially the force peak.

Quantization resolution The quantization resolution mainly determines the achievable dynamic range D, defined as the ratio between the highest and lowest values representable after digitization. It can be approximated by $D \approx 6 \cdot n_{\mathrm{ADC}}$ and is usually around 90–100 dB for most vibration measurement transducers [2]. Considering the cost constraint, the lower bound of 90 dB is set as a requirement. This results in a minimum ADC resolution n_{ADC} of 15 bits. Since the usual measurement signals, e.g., accelerations, also carry a sign, a bipolar ADC with 16-bit resolution is the minimum.

Time synchronicity In order to quantify the errors resulting from synchronicity issues, a simple numerical model is set up, namely a lumped mass oscillator with five degrees of freedom (dof) as depicted in Fig. 9.3.

The model consists of five masses, connected with spring–damper elements, with the leftmost one rigidly connected to the environment. The matrices are set up in Python and integrated using Newmark's method with $\gamma = \frac{1}{2}$ and $\beta = \frac{1}{4}$ (average constant acceleration) [4], using the identified, minimum sampling rate of 10 kHz. The system is excited by an impulse-shaped cosine that is applied successively to each dof. A start time offset Δt_{ST} or drift per second τ_{drift} can then be applied to the displacement signals. In all cases, the full frequency response function (FRF) matrix is calculated using *pyFRF* [5]. The modal parameters and eigenvectors are then extracted with *pyFBS* [6], using a combination of the poly-reference Least-Squares Complex Frequency (pLSCF) and Least-Squares Frequency Domain (LSFD) methods [7]. The original, i.e., unmodified, modes are shown in Fig. 9.4.

Figure 9.4 shows the real part of the eigenvectors, after they have been normalized such that the highest amplitude is one and rotated such that dof 1 has no imaginary part. While this might be handy for the visualization of real modes, it also removes information about the phase between the different nodes. Since the used model is proportionally damped (damping matrix can be constructed from a linear combination of the mass and stiffness matrices), the eigenvectors are purely real and the phase is either 0° or 180° [8]. But having one (sensor) signal shifted in time due to a start time offset Δt_{ST} introduces a phase shift depended on frequency. Such (small) changes in phase are then not clearly visible, as illustrated in Fig. 9.5.

Phase perturbations of purely real eigenvectors by small values are barely visible when only considering the real part. They are better visible after a rotation ωt, with ω being the eigenfrequency corresponding to the eigenvector, but the amplitude is arbitrary. To tackle this issue, the eigenvectors are shown in the complex plane. Since the phase shift due to start time offsets Δt_{ST} is increasing with frequency, the mode with the highest eigenvalue is affected the most. Therefore, only mode

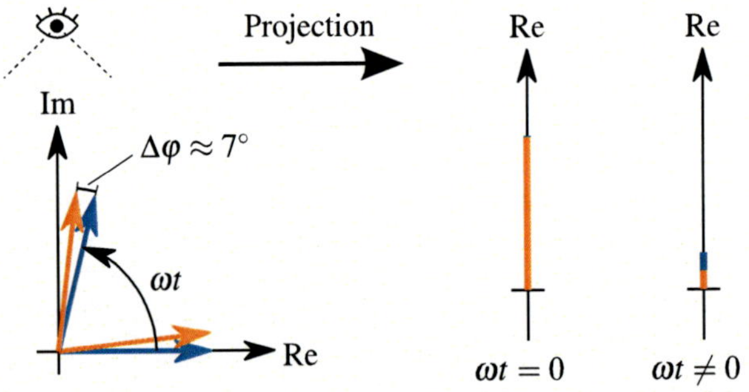

Fig. 9.5 Illustration of issue arising from viewing only the real part of eigenvectors: small phase changes of (mostly) real eigenvectors are barely recognizable, blue: unmodified and orange: phase shift of $\Delta\varphi \approx 7°$ introduced

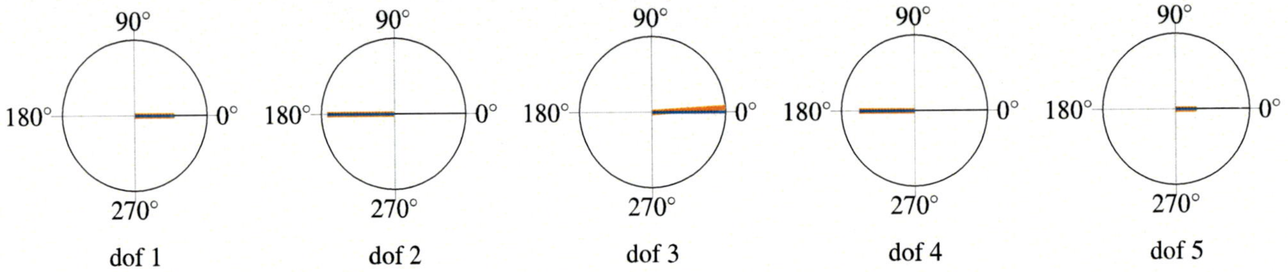

Fig. 9.6 Visualization of mode 5 in the complex plane, with dof 3 shifted in time by one sample (orange), $\Delta\varphi \approx 3.5°$

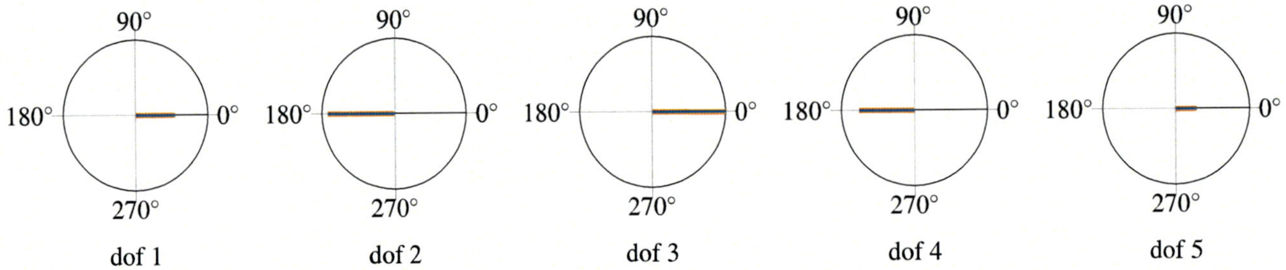

Fig. 9.7 Visualization of mode 5 in the complex plane, with dof 3 drifting 30 µs/s, $\Delta\varphi \approx -0.1°$

5 is shown in the following. The synchronicity issues are only applied to dof 3, to simulate one board not being perfectly synchronous.

At first, the influence of a start time offset Δt_{ST} is investigated. In order to avoid a resampling of the signal, the shifting is done in steps of $\Delta t = 1/f_S$, resulting in the smallest offset being 100 µs. The components of the fifth eigenvector are shown in Fig. 9.6, with the original eigenvector components in blue and the ones with start time offset Δt_{ST} applied to dof 3 in orange.

As can be seen, the dof 3 component now has an imaginary part or in other words is rotated in the complex plane by about 3.5°. The same value can be found by calculating the phase shift with $\Delta\varphi = \Delta t_{\mathrm{ST}}/T_{\mathrm{mode}} \cdot 360°$, where T_{mode} is the period of the respective mode. Calculating the phase shift of a theoretical mode at 3 kHz, assuming the relation holds, results in an error of 108°, which is by no means acceptable. A maximum phase shift error of 10° at the highest frequency of interest is set as the limit requirement. This results in a maximally allowed start time offset Δt_{ST} of about 9 µs. This requirement is automatically fulfilled within one system with a simultaneously sampling ADC, but is very challenging for connecting multiple devices.

The second issue of drift between clocks is investigated based on the observed drift of two *Espressif ESP32* microcontrollers. Both controllers were sampling the output of a signal generator for 500 s, after which the offset between the sampled signal and a generated, perfect signal was compared. They both showed a drift τ_{drift} of 30 µs/s. A simulation of this kind of drift results in a neglectable phase error of −0.1°, which is not even visible in the complex plane, see Fig. 9.7.

This error is smaller than the one resulting from a start time offset $\Delta t_{\mathrm{ST}} \leqslant 9$ µs, meaning that the error introduced by drifting over time is negligible in comparison. Therefore, the determined *Espressif ESP32* performance is set as the minimum requirement. Table 9.1 summarizes the determined performance requirements.

Table 9.1 Summary of the identified minimum requirements for data acquisition for experimental vibration measurements

Quantity	Minimum requirement
Sample rate f_S	$\geqslant$10 kHz
ADC resolution n_{ADC}	$\geqslant$16 bit
Start time offset Δt_{ST}	$\leqslant$9 µs
Drift τ_{drift}	$\leqslant$30 µs/s

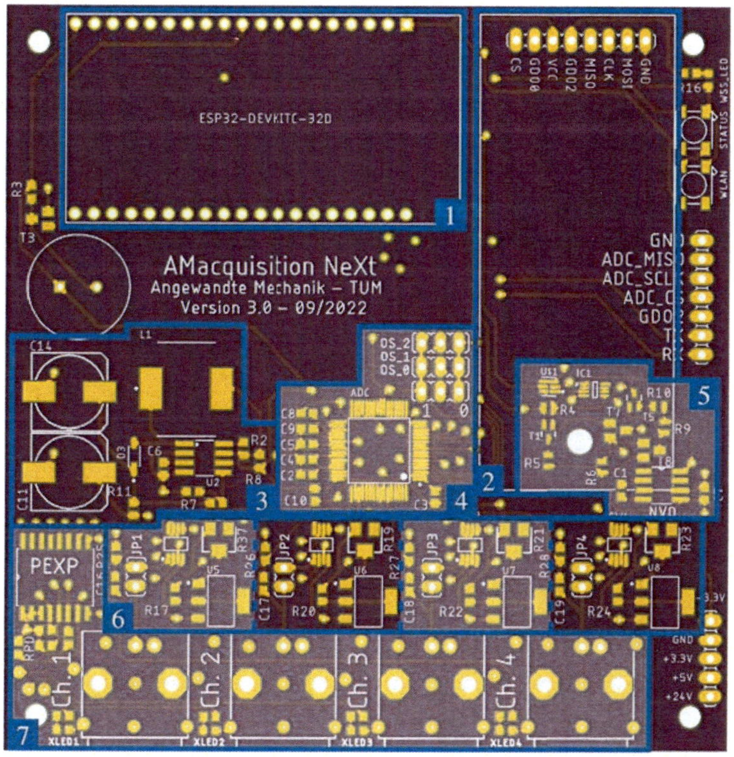

Acquisition board sub-modules

1 Micro-controller module
2 Wireless synchronization module
3 24 V boost converter
4 Analog-Digital-Converter (ADC)
5 Logic inverter for TEDS readout
6 Acquisition back end (1 per channel)
7 Acquisition front end

Fig. 9.8 Overview of the designed acquisition board with important sub-modules highlighted and described below

9.3 Acquisition Board Design

The requirements derived in the previous section were incorporated into a board design that was further refined and extended over two iterations. The built acquisition system offers four channels for IEPE signals, which can be sampled with up to 20 kHz and 16-bit resolution continuously. Making use of the PSRAM up to 50 kHz can be achieved within a limited period. Additionally, the readout of the transducer electronic data sheets (TEDSs) is possible. Thanks to the micro-controller's Wi-Fi capability, both wired (single USB 2.0 connection) and wireless operation modes are available. Figure 9.8 shows the current design and highlights important sub-modules, whose functions will be described in the following. As part of the open-source requirement, full schematics and software can be found on the official page: https://github.com/Official-OASIS-Project.

Micro-controller module The self-contained *Espressif Systems ESP32-DevKitC-VE* development kit is used, which incorporates all required components for the micro-controller, i.e., a 3.3 V supply or a USB-to-UART Bridge supporting up to 3 Mbps. The latter is important to achieve a sample rate of 20 kHz with a wired connection, and other kits with 1 Mbps transfer rate are not sufficient. Another reason for this specific kit is the *Espressif Systems ESP32-WROVER-E* module, which possesses additional PSRAM used for caching acquisition data. Since most computation time for processing a sample is spent on the data transfer from the micro-controller to an external device, moving this step after the actual acquisition phase greatly increases the achievable sample rate to up to 50 kHz. The built-in Wi-Fi module additionally enables to retrieve the acquisition data wirelessly.

Wireless synchronization module This module sends and receives synchronization impulses to or from other acquisition boards. It is configured by the micro-controller via SPI and provides a rising edge when a synchronization impulse was successfully sent or received. The design is based on the *Texas Instruments CC1101* chip and implements the reference design.

Analog–Digital Converter (ADC) According to the derived requirements, a simultaneously sampling, bipolar ADC was selected. With the overall budget limit in mind, the cheaper *AD7606-4* from *Analog Devices* was selected. Compared to the more than twice as expensive flagship *AD7606C-18*, it lacks 2 bits of resolution (16 bits instead of 18 bits) and the ability to configure it via SPI. Therefore, access to the oversampling configuration is provided with jumpers. These two ADC models are not pin compatible and require further changes if an interchange is desired.

Logic inverter for TEDS readout In order to read data from the sensor EEPROM, a signal with negative potential with respect to ground has to be applied to the TEDS enabled sensor. Data can be read using the OneWire protocol and the standard library using this logic inverter circuit. So far the specifics for the communication with accelerometers using TEDS version 0.9 have been worked out and successfully validated separately. At the time of writing, the implementation of the logic inverter shows instability issues and cannot be used reliably.

24 V boost converter IEPE sensors require a constant current source that can provide up to 24 V. This module creates a 24 V voltage source from the 5 V provided by the USB connection. This is done with a boost converter using the *MC34063A* chip, implementing the reference design with components chosen such that 25 mA can be provided continuously. An additional RC filter is fitted to reduce the voltage ripple.

Acquisition back end Each channel has its own back end consisting of a constant current source provided by the *Analog Devices LT3092* chip. The current is adjustable using a potentiometer. The back end also handles the switching between current source and TEDS logic inverter using the *Analog Devices ADG419* chip. Furthermore, a high-pass filter is implemented that removes the DC part (commonly near 12 V) of the sensor signal. The filter is designed such that the influence on frequencies above 10 Hz is minimal, not disturbing the usual working range of piezo accelerometers.

Acquisition front end The front end contains the BNC connectors for the sensors as well as a port expander (*PCA9554A*), which controls the switching of the *ADG419* chips as well as LEDs for status feedback. Using an expander is necessary as there are not enough channels left on the micro-controller otherwise. Additional LEDs and a buzzer are fitted on the board for extensive feedback for the user, which also proved quite helpful during troubleshooting, just as the optional debug headers.

9.4 Wireless Sample Synchronization: Operation Principle and Validation

The idea is to have one device act as the source, which transmits a synchronization impulse, and arbitrary many secondary devices, the sinks, which wait for an impulse to be sent. Sending and receiving of said impulse is handled by the *CC1101* chip that handily provides a rising edge after sending or receiving a sync word, which is used to create a GPIO interrupt, see also Fig. 9.9.

Based on oscilloscope measurements, the dead time t_{dead} between two boards due to the *CC1101* chip is 25.94 µs±0.05 µs (95 % confidence) and can therefore be partially corrected in software. The boards are programmed to start sampling when the GPIO interrupt is serviced by the micro-controller. Since the derivation of the minimum requirements showed that the drift of the *ESP32* clocks is neglectable, only the start of the sampling is synchronized, and there is no re-synchronization during the sampling. In order to validate this concept, two tests are performed: a synthetic one, where a step function from a signal generator is sampled, and an experimental one, measuring vibrations on a benchmark structure.

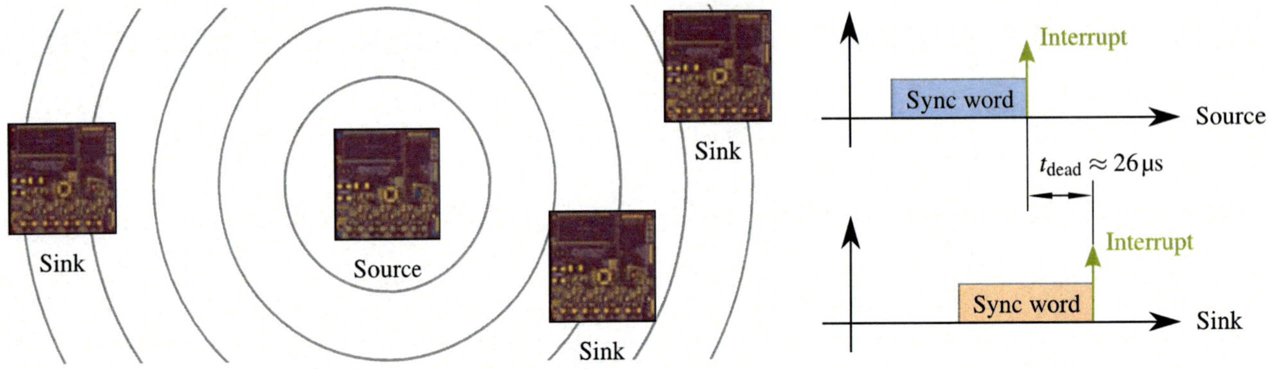

Fig. 9.9 Concept for wireless sample synchronization, including timing diagram

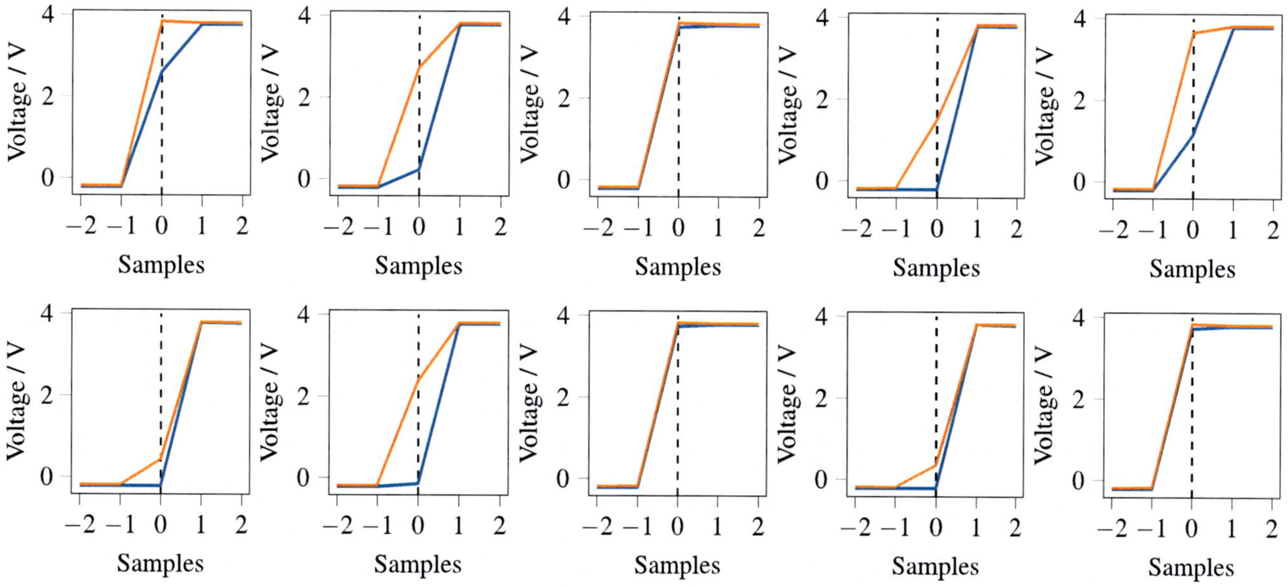

Fig. 9.10 Plot of synthetic validation series, measuring step functions with 50 ns rise time, plotted over samples such that first observable change in any signal happens at sample number 0, blue: board one and orange: board two, $f_S = 10\,\mathrm{kHz}$

The signal used for the synthetic test is a step function with an amplitude of 4 V. Using an oscilloscope, the rise time of the signal is identified to be approximately 50 ns. One pair of signals (one source, one sink) is aligned such that the first change in any signal occurs at sample number 0. A total of 10 tests were performed with $f_S = 10\,\mathrm{kHz}$, whose results are shown in Fig. 9.10.

At a first glance, the results look like the start time offset Δt_{ST} might be smaller than the rise time of 50 ns, when looking at plots 2, 5, and 7, where both signals have neither of the extreme values at sample number 0. But this is not the case and the reason is found in the oversampling, which is enabled to reduce noise. With the configuration used, each recorded sample is the average of 8 samples taken by the ADC. Those additional samples are acquired with an ADC internal sample rate of approximately 200 kHz. While it might be possible to calculate an estimate for the delay between the two signals from the afore mentioned or using statistical methods, this is not the goal here. Based on the time the ADC requires to internally gather 8 samples, roughly 35 µs, the delay between the two boards is well below 1 sample (i.e., 100 µs).

For the second validation, the vibrations of a benchmark structure, an L-shaped aluminum beam welded to a plate, are measured. One impact hammer and 3 one-axis acceleration sensors attached to the top of the L-beam are used. The signals are split using tee connectors, so both boards receive the identical analog signal, see Fig. 9.11.

In order to determine the performance of the wireless sample synchronization, the cross-correlation between the identical analog signals captured by the two wirelessly synchronized boards is calculated. For comparison reasons, the auto-correlation of one board is shown in the plots in Fig. 9.12 as well. The correlations are shown in dependence of the offset τ in samples, with the highest amplitude normalized to one.

All cross-correlations have their maximum at $\tau = 0$ but appear to be slightly shifted toward the right. The goal is to only get a rough idea of the expectable delay for now. Just as with the synthetic test, the delay can be said to be well below one sample or 100 µs, but it most likely does not meet the set requirement yet.

9.5 Conclusion

In the beginning, the minimum requirements for experimental vibration measurements in an academic context were derived. These requirements could mostly be met by the developed and described acquisition system. The proposed synchronization approach using the *Texas Instruments CC1101* chip works as intended and is able to synchronize two or more boards with less than 100 µs of delay. Nonetheless, further fine-tuning of the data acquisition software and validation techniques is required.

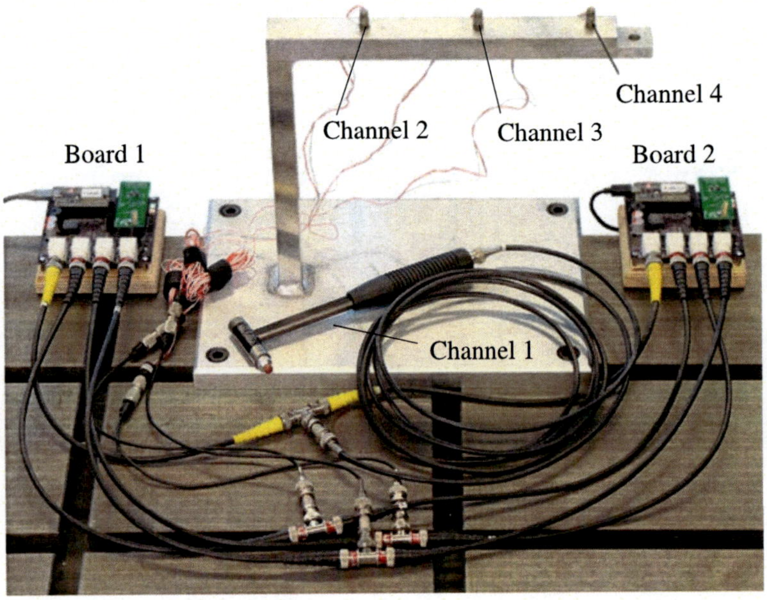

Fig. 9.11 Experimental setup for validation: one impact hammer as excitation and 3 one-axis acceleration sensors attached to the benchmark structure, signals are split using BNC cables and fed into both acquisition boards

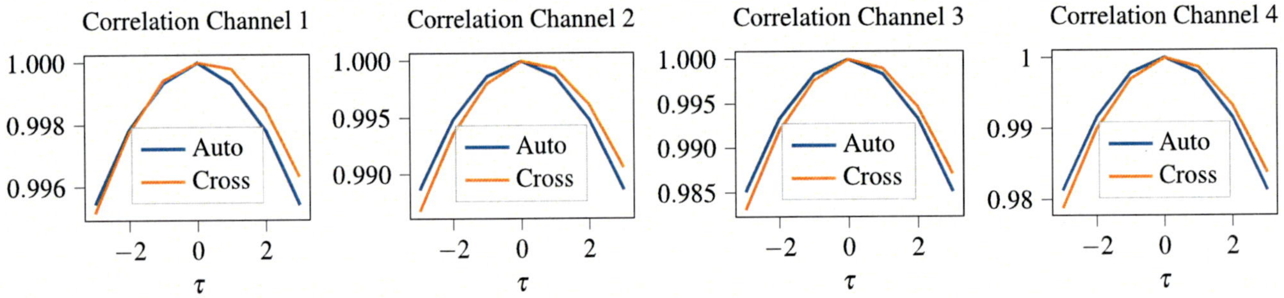

Fig. 9.12 Auto- and cross-correlation of experimental validation for each channel. Auto-correlation calculated from signals of board 1, cross-correlation between boards 1 and 2, highest amplitude normalized to 1, offset τ in samples with $1 \cdot \tau = 100\,\mu s$

With the methods used so far, an exact determination of the delay between two boards is not possible. This inhibits both the identification of the exact delay and the variance. While this approach might not be as precise as commercial products, it still enables measurements where a synchronization by cable is not possible and external signals, like GPS, cannot be used. As the schematics and software are made publicly available, this enables insight into the working principles and allows everyone to expand upon this basis.

References

1. León, F.P.: Messtechnik: Grundlagen, Methoden und Anwendungen. Springer Vieweg Berlin, Heidelberg (2019)
2. Brandt, A.: Noise and Vibration Analysis: Signal Analysis and Experimental Procedures, 1st edn. Wiley, Chichester (2011)
3. Trainotti, F., Berninger, T., Rixen, D.: Using Laser Vibrometry for Precise FRF Measurements in Experimental Substructuring (2020)
4. Géradin, M., Rixen, D.J.: Mechanical Vibrations: Theory and Application to Structural Dynamics, 3rd edn. Wiley, Chichester (2014)
5. Open Modal Project. pyFRF (v0.40) (2019)
6. Bregar, T., Mahmoudi, A.E., Kodrič, M., Ocepek, D., Trainotti, F., Pogačar, M., Göldeli, M., Čepon, G., Boltežar, M., Rixen, D.J.: pyFBS: a python package for frequency based substructuring (v0.2.9). J. Open Source Softw. **7**(69), 3399 (2022)
7. Guillaume, P., Verboven, P., Vanlanduit, S., Van der Auweraer, H., Peeters, B.: A poly-reference implementation of the least-squares complex frequency-domain estimator. Proc. IMAC **21**, 01 (2003)
8. Ewins, D.J.: Modal Testing: Theory, Practice and Application. Wiley, Chichester (2000)

Chapter 10
Design of a Variable Stiffness Impact Damper Using Magnetorheological Elastomers

Diego Francisco Ledezma-Ramírez, Emiliano Rustighi, and Pablo Ernesto Tapia-González

Abstract The suppression and control of mechanical vibrations, shock, and impacts is a subject of great interest for different areas of engineering, due to their potential negative effects, such as noise, fatigue, and mechanical failure in general. This is particularly important for the aerospace and automotive industries where in addition weight reduction is paramount. Usually, isolation and suppression methods are used either in the form of resilient isolators intended to absorb and dissipate the energy or in the form of tuned mass dampers. However, recently the applications where nonlinear elements are implemented have become more relevant. A particular nonlinear strategy for vibration and shock suppression is the impact dampers. These devices work on the principle of transferring momentum from the vibrating structure to the damper through impacts between them. There are several proposed designs of impact dampers, either configured as an end stop, also called vibro-impact attachments, or particle impact dampers, where small particles are contained into a vibrating structure or attached to it, and the collisions between the particles and the confinement result in energy dissipation and thus vibration suppression. In this work, the design of a variable stiffness impact damper is proposed. By controlling the stiffness of the damper, the amount of energy absorption and dissipation can be enhanced during impacts. To build the impact damper, a magnetorheological elastomer (MRE) is considered. These elastomers are manufactured using silicone rubber with embedded ferromagnetic particles, achieving stiffness variation through the application of a magnetic field. The design and modeling of the damper using a nonlinear geometry is introduced, and then prototypes are manufactured and experimentally quantified in terms of their capacity to change stiffness when a magnetic field is applied, as well as the energy absorption and vibration suppression capabilities. The advantages and limitation of the proposed design are then presented and discussed. It is found how the impact damper with variable stiffness can improve the vibration suppression under certain conditions depending on the level of the impact and the stiffness change obtained. Discussion about further control strategies and implementation are presented.

Keywords Impact damper · Variable stiffness · Shock · Vibration isolation · Active control

10.1 Introduction

Recently, there has been an increased interest in the application of nonlinear passive, active, or adaptive control and isolation strategies for vibration, shock, and impacts [1]. An idea for impact suppression is to transfer the energy resulting from a shock to a secondary system rather than isolating the vibration. Karayannis et al. [2] presented the concept of vibro-impact attachments investigating their capacity to passively absorb and dissipate significant portions of shock energy applied to the primary systems finding improved isolation under a wide frequency range. A device that transfers part of the momentum to a damper during an impact thus reducing shock vibration was presented by Son et al. [3]. Active and semiactive strategies have also been adapted to the later concept, i.e., by using an actuator to change the stiffness force and compensate the shock applied to the main mass [4]. Magnetorheological elastomers (MREs), which are variable stiffness materials obtained by mixing a matrix such as silicone rubber and magnetic iron particles, have also been used as impact dampers in [5] where the objective was to isolate an impact by controlling the effective damping of the system. In this work, a simple compressional

D. F. Ledezma-Ramírez (✉) · P. E. Tapia-González
Facultad de Ingenieria Mecánica y Eléctrica, Universidad Autónoma de Nuevo León, Nuevo León, México
e-mail: diego.ledezmard@uanl.edu.mx; pablo.tapiagz@uanl.edu.mx

E. Rustighi
Industrial Engineering Department, University of Trento, Trento, Italy
e-mail: emiliano.rustighi@unitn.it

C. Walber et al. (eds.), *Sensors & Instrumentation and Aircraft/Aerospace Testing Techniques, Volume 8*, Conference Proceedings of the Society for Experimental Mechanics Series, https://doi.org/10.1007/978-3-031-34938-6_10

MRE is evaluated as an impact damper. The design of the proposed damper comprises an impacting mass attached to a cantilever beam which is subjected to a sudden release which then impacts the MRE damper. Different configurations of the magnetic field supplied to the damper are evaluated to investigate the effect on the impact suppression.

10.2 Background

The MRE damper was manufactured using silicone rubber (MM288 by ACC Silicones Ltd) as matrix and carbonyl iron powder (CIP Type SQ by BASF SE) as ferromagnetic particles with an average diameter of 5 μm . These properties result in high magneto-rheological effects. Two dampers with particle concentration of 0 and 30% were manufactured in a cylindrical shape of 30 mm diameter and a 12 mm height. The rubber was cured at the room temperature for 24 h without the effect of a magnetic field. The samples are designed to work in axial load, and when a magnetic field is applied in this direction, a change in stiffness of about 40% can be obtained depending on the magnitude of the magnetic field and the configuration, and the stiffness can be reduced or increased from a baseline depending on the polarity of the magnetic field. The details of the properties, modeling, and change in stiffness under an applied magnetic field can be found in reference [6] (Fig. 10.1).

The proposed experimental device comprises an aluminum cantilever beam with an impacting mass in the free end. The MRE impact damper is attached to an electromagnet which is supported in a fixed frame. There is a gap of 3 mm between the impact damper and the mass on the beam. A mechanical impedance sensor between the frame and the electromagnet is used to record the transmitted force and acceleration during an impact, and an additional accelerometer is placed on the tip of the beam to record its acceleration. By means of a sudden release of the beam, the mass impacts the damper.

10.3 Experimental Results

For the experiments, the cantilever beam was pulled a fixed distance of 6 mm far from the impact damper and then suddenly released. The transmitted force and acceleration were recorded and compared for different conditions of the impact damper, i.e., the electromagnet was first turned off, then it was turned on operating at 20 VDC, and finally the polarity was reversed maintaining the same voltage. In addition, an impact damper with no CIP particles, i.e., 0% of concentration, was considered as well. Figure 10.2 shows the transmitted force for all the cases discussed. Figure 10.3 presents the comparison between the input acceleration measured at the tip of the beam and the transmitted acceleration through the impact damper.

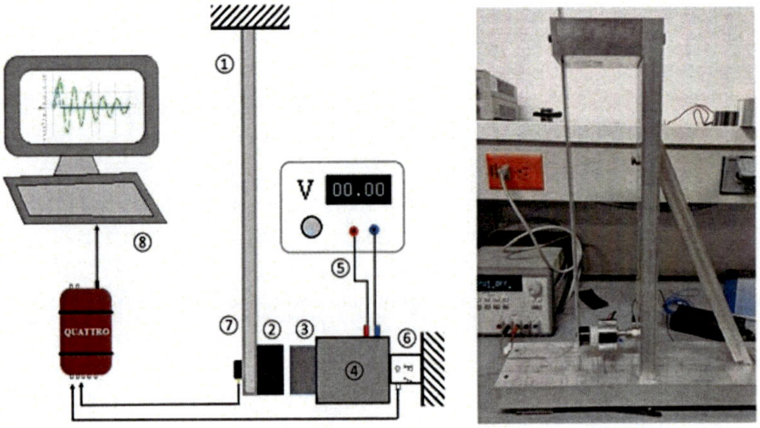

Fig. 10.1 Schematics and picture of the experimental rig. (1) Cantilever beam, (2) impact mass, (3) MRE impact damper, (4) DC electromagnet, (5) DC power supply, (6) impedance sensor, (7) accelerometer, and (8) PC-based data acquisition system

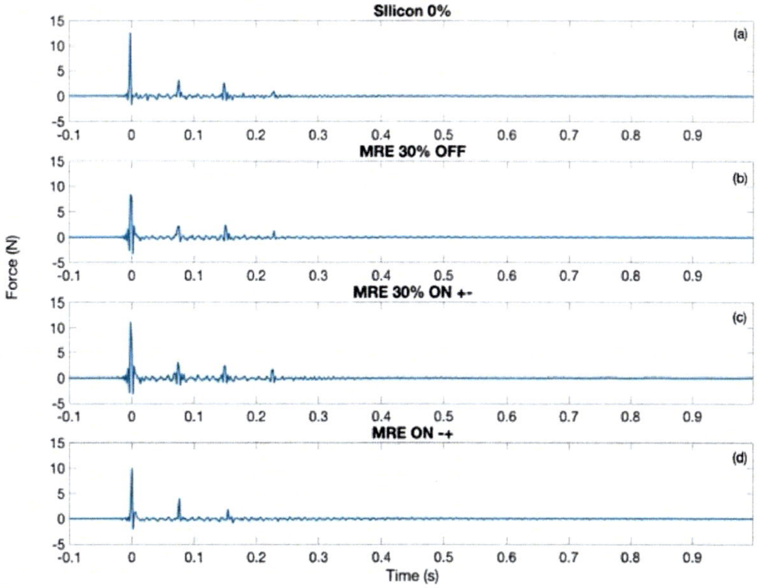

Fig. 10.2 Transmitted force due to mass impacting the MRE damper for the different cases considered. (**a**) 0% damper, (**b**) MRE damper OFF, (**c**) MRE damper ON, and (**d**) MRE ON inverted voltage

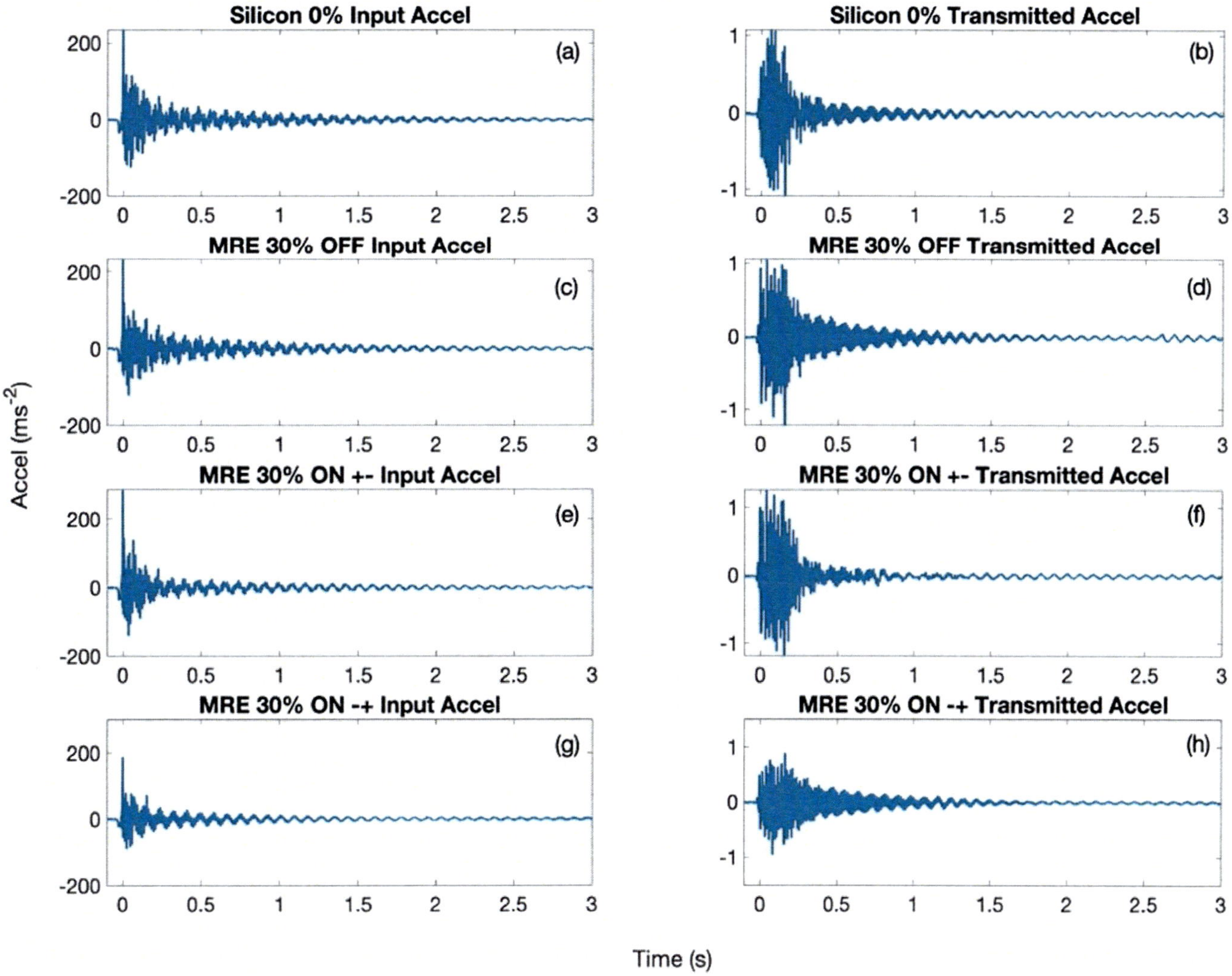

Fig. 10.3 Input and transmitted acceleration due to mass impacting the MRE damper for the different cases considered. (**a**, **b**) 0% damper, (**c**, **d**) MRE damper OFF, (**e**, **f**) MRE damper ON, and (**g**, **h**) MRE ON inverted voltage

10.4 Conclusion

The design of an MRE impact damper was presented and demonstrated experimentally, using a cantilever beam to provide an impact to the damper by means of an attached mass to the free end of the beam. Two impact dampers were studied, i.e., one with no particles and a sample with particle concentration of 30% under different conditions of the applied magnetic field to study the effect of changing the stiffness during an impact. By using MRE, it is possible to control the stiffness of the impact damper and then reduce the amount of force and acceleration transmitted through the damper. The idea presented allows for further research into the active control of the damper where the implementation of a control strategy could lead to further improvements in impact and vibration suppression.

References

1. Ledezma-Ramírez, D.F., Tapia-González, P.E., Ferguson, N., Brennan, M., Tang, B.: Recent advances in shock vibration isolation: an overview and future possibilities. Appl. Mech. Rev. **71**(6), 060802 (2019)
2. Karayannis, I., Vakakis, A.F., Georgiades, F.: Vibro-impact attachments as shock absorbers. Proc. Inst. Mech. Eng. Part C: J. Mech. Eng. Sci. **222**(10), 1899–1908 (2008)
3. Son, L., Kawachi, M., Matsuhisa, H., Utsuno, H.: Reducing floor impact vibration and sound using a momentum exchange impact damper. J. Syst. Des. Dyn. **1**(1), 14–26 (2007)
4. Son, L., Hara, S., Yamada, K., Matsuhisa, H.: Experiment of shock vibration control using active momentum exchange impact damper. J. Vib. Control **16**(1), 49–64 (2010)
5. Rahmat, M.S., Hudha, K., Kadir, Z.A., Nuri, N.R.M., Amer, N.H., Abdullah, S.: Modelling and control of a magneto-rheological elastomer for impact reduction. J. Mech. Eng. Sci. **13**(3), 5259–5277 (2019)
6. Rustighi, E., Ledezma-Ramirez, D.F., Tapia-Gonzalez, P.E., Ferguson, N., Zakaria, A.: Modelling and experimental characterisation of a compressional adaptive magnetorheological elastomer isolator. J. Vib. Control **28**(21–22), 3093–3107 (2022)

Chapter 11
Realization of a Virtual Acoustic Black Hole with Piezoelectric Patches

Samuel Quaegebeur, Ghislain Raze, Li Cheng, and Gaëtan Kerschen

Abstract Reducing the mechanical vibration is of the utmost importance to lower mechanical stress and thus extend the life of a structure. This work proposes a novel concept to achieve this through an acoustic black hole (ABH) effect implemented via a digital controller. An ABH is a device that localizes the vibrational energy, which is in turn dissipated using damping layers. Its practical realization consists of a tapered wedge beam whose thickness follows a power-law profile. Its efficiency usually starts beyond a cut-on frequency, which is inversely proportional to its length. Obtaining the ABH effect on slender structures is thus very challenging: to achieve vibration reduction at low frequencies, the tapered wedge beam must be very long and thin. We propose herein to circumvent this problem by using a digital controller connected to piezoelectric transducers which are bonded to the host structure. Digital controllers have the significant advantage of being able to reproduce virtually any desired mechanical impedance function and, in particular, that of an ABH. We verify the soundness of the approach through detailed numerical simulations. Those are conducted on a one-dimensional slender beam modeled by the finite element method. The simulations show promising results, and the practical realization of the virtual acoustic black hole (VABH) is discussed eventually.

Keywords Acoustic black hole · Active control · Vibration reduction · Piezoelectric · Dynamic substructuring

11.1 Introduction

The acoustic black hole effect [1] has been the subject of many innovations for the past two decades [2]. This device is obtained by changing the thickness of a profile into a power-law function. The vibrational energy then gets localized inside it, and different strategies can be employed to mitigate the vibrations [3, 4]. However, the ABH effect is efficient above a cut-on frequency which is inversely proportional to the length of the profile. To reduce vibration at low frequencies, very long and thus thin profiles must be manufactured which is very challenging. Some strategies [5, 6] have been developed over the years to enhance the ABH effect without modifying its profile. However, the efficiency of such approaches remains limited and is not as good as if one could create a very long and thin profile.

Piezoelectric shunts [7, 8] correspond to another approach to mitigate vibrations. Attaching a piezoelectric patch to a host enables to convert mechanical energy into electrical energy. Connecting the patch to a shunt (for instance, a resonant shunt) enables to dissipate the energy and thus to reduce the vibrations. However, the components forming the shunt may be difficult to create in practice or ill-adapted. Digital vibration absorbers (DVAs) [9] allow to circumvent this problem. Instead of plugging a passive electrical circuit to the piezoelectric patches, it is connected to a digital unit and a source of power to reproduce any electrical impedance function. These systems are thus very versatile and are used for a wide range of applications such as nonlinear vibration [10].

The key idea of this research is to propose a physical implementation of a VABH. Such a concept enables to tackle any practical issues encountered with ABH and to create designs that could never be obtained in practice.

S. Quaegebeur (✉) · G. Raze · G. Kerschen
Aerospace and Mechanical Engineering Department, University of Liège, Liège, Belgium
e-mail: squaegebeur@uliege.be; g.raze@uliege.be; g.kerschen@uliege.be

L. Cheng
Department of Mechanical Engineering, The Hong Kong Polytechnic University Hung Hom, Kowloon, Hong Kong SAR, China
e-mail: li.cheng@polyu.edu.hk

C. Walber et al. (eds.), *Sensors & Instrumentation and Aircraft/Aerospace Testing Techniques, Volume 8*, Conference Proceedings of the Society for Experimental Mechanics Series, https://doi.org/10.1007/978-3-031-34938-6_11

11.2 Background

Figure 11.1 presents the mechanical structure under study. All parameters of the system are provided in Table 11.1. While the uniform beam has a constant thickness h_0, the tapered wedge beam's profile satisfies the following law:

$$h(x) = h_0 \left(\frac{L + L_{\mathrm{ABH}} - x}{L_{\mathrm{ABH}}} \right)^m, \quad x \in [L, L + L_{\mathrm{ABH}} - x_0], \tag{11.1}$$

where x_0 denotes the tapered wedge beam truncation. The beam is clamped at its extremity $x = 0$ and free at the tip. The cut-on frequency of the device [11] is equal to 40.9 Hz.

The system is discretized with the finite element method. Each node is characterized by three displacements: the longitudinal and transversal displacements and the rotational displacement. Superscripts b and tb denote the uniform and tapered wedge beams, respectively. For both systems, the internal nodes are noted by the subscript I and the boundary nodes (nodes shared by both beams) by B. The equations of motion of both systems read

$$\begin{bmatrix} \mathbf{M}^{\mathrm{b}}_{\mathrm{II}} & \mathbf{M}^{\mathrm{b}}_{\mathrm{IB}} \\ \mathbf{M}^{\mathrm{b}}_{\mathrm{BI}} & \mathbf{M}^{\mathrm{b}}_{\mathrm{BB}} \end{bmatrix} \begin{bmatrix} \ddot{\mathbf{x}}^{\mathrm{b}}_{\mathrm{I}} \\ \ddot{\mathbf{x}}_{\mathrm{B}} \end{bmatrix} + \begin{bmatrix} \mathbf{C}^{\mathrm{b}}_{\mathrm{II}} & \mathbf{C}^{\mathrm{b}}_{\mathrm{IB}} \\ \mathbf{C}^{\mathrm{b}}_{\mathrm{BI}} & \mathbf{C}^{\mathrm{b}}_{\mathrm{BB}} \end{bmatrix} \begin{bmatrix} \dot{\mathbf{x}}^{\mathrm{b}}_{\mathrm{I}} \\ \dot{\mathbf{x}}_{\mathrm{B}} \end{bmatrix} + \begin{bmatrix} \mathbf{K}^{\mathrm{b}}_{\mathrm{II}} & \mathbf{K}^{\mathrm{b}}_{\mathrm{IB}} \\ \mathbf{K}^{\mathrm{b}}_{\mathrm{BI}} & \mathbf{K}^{\mathrm{b}}_{\mathrm{BB}} \end{bmatrix} \begin{bmatrix} \mathbf{x}^{\mathrm{b}}_{\mathrm{I}} \\ \mathbf{x}_{\mathrm{B}} \end{bmatrix} = \begin{bmatrix} \mathbf{f}_{\mathrm{ext,I}} \\ \mathbf{f}_{\mathrm{ext,B}} + \mathbf{f}_{\mathrm{tb}\to\mathrm{b}} \end{bmatrix} \tag{11.2a}$$

$$\begin{bmatrix} \mathbf{M}^{\mathrm{tb}}_{\mathrm{II}} & \mathbf{M}^{\mathrm{tb}}_{\mathrm{IB}} \\ \mathbf{M}^{\mathrm{tb}}_{\mathrm{BI}} & \mathbf{M}^{\mathrm{tb}}_{\mathrm{BB}} \end{bmatrix} \begin{bmatrix} \ddot{\mathbf{x}}^{\mathrm{tb}}_{\mathrm{I}} \\ \ddot{\mathbf{x}}_{\mathrm{B}} \end{bmatrix} + \begin{bmatrix} \mathbf{C}^{\mathrm{tb}}_{\mathrm{II}} & \mathbf{C}^{\mathrm{tb}}_{\mathrm{IB}} \\ \mathbf{C}^{\mathrm{tb}}_{\mathrm{BI}} & \mathbf{C}^{\mathrm{tb}}_{\mathrm{BB}} \end{bmatrix} \begin{bmatrix} \dot{\mathbf{x}}^{\mathrm{tb}}_{\mathrm{I}} \\ \dot{\mathbf{x}}_{\mathrm{B}} \end{bmatrix} + \begin{bmatrix} \mathbf{K}^{\mathrm{tb}}_{\mathrm{II}} & \mathbf{K}^{\mathrm{tb}}_{\mathrm{IB}} \\ \mathbf{K}^{\mathrm{tb}}_{\mathrm{BI}} & \mathbf{K}^{\mathrm{tb}}_{\mathrm{BB}} \end{bmatrix} \begin{bmatrix} \mathbf{x}^{\mathrm{tb}}_{\mathrm{I}} \\ \mathbf{x}_{\mathrm{B}} \end{bmatrix} = \begin{bmatrix} \mathbf{0} \\ -\mathbf{f}_{\mathrm{tb}\to\mathrm{b}} \end{bmatrix}, \tag{11.2b}$$

where **M**, **C**, and **K** correspond to the mass, damping, and stiffness matrices. External forces, $\mathbf{f}_{\mathrm{ext}}$, are applied to the uniform beam, but we assume that none of them are applied to the tapered wedge beam. Both beams share a boundary where internal forces, $\mathbf{f}_{\mathrm{tb}\to\mathrm{b}}$, are transmitted. This term translates the ABH effect on the uniform beam. Using the Laplace transform, with s its variable, in Equation (11.2b), gives

$$\mathbf{X}_{\mathrm{I}} = \underbrace{\left(s^2\mathbf{M}^{\mathrm{tb}}_{\mathrm{II}} + s\mathbf{C}^{\mathrm{tb}}_{\mathrm{II}} + \mathbf{K}^{\mathrm{tb}}_{\mathrm{II}}\right)^{-1} \left(s^2\mathbf{M}^{\mathrm{tb}}_{\mathrm{IB}} + s\mathbf{C}^{\mathrm{tb}}_{\mathrm{IB}} + \mathbf{K}^{\mathrm{tb}}_{\mathrm{IB}}\right)}_{\mathbf{Z}_{\mathrm{IB}}(s)} \mathbf{X}_{\mathrm{B}} \tag{11.3a}$$

$$-\mathbf{F}_{\mathrm{tb}\to\mathrm{b}} = \underbrace{\left[\left(s^2\mathbf{M}^{\mathrm{tb}}_{\mathrm{BI}} + s\mathbf{C}^{\mathrm{tb}}_{\mathrm{BI}} + \mathbf{K}^{\mathrm{tb}}_{\mathrm{BI}}\right) \mathbf{Z}_{\mathrm{IB}}(s) + \left(s^2\mathbf{M}^{\mathrm{tb}}_{\mathrm{BB}} + s\mathbf{C}^{\mathrm{tb}}_{\mathrm{BB}} + \mathbf{K}^{\mathrm{tb}}_{\mathrm{BB}}\right)\right]}_{\mathbf{Z}_{\mathrm{ABH}}} \mathbf{X}_{\mathrm{B}}, \tag{11.3b}$$

where $\mathbf{X}_{\mathrm{I}}$, $\mathbf{X}_{\mathrm{B}}$, and $\mathbf{F}_{\mathrm{tb}\to\mathrm{b}}$ are the Laplace transforms of $\mathbf{x}^{\mathrm{tb}}_{\mathrm{I}}$, $\mathbf{x}_{\mathrm{B}}$, and $\mathbf{f}_{\mathrm{tb}\to\mathrm{b}}$.

For fast computation and efficiency, it is convenient to reduce the finite element models with a Craig–Bampton reduction [12]. Boundary nodes are kept as master nodes, and a sufficient number of modes are kept in the basis to have very good

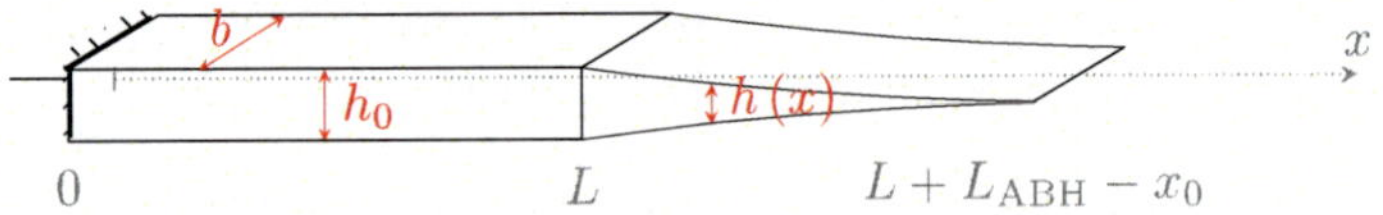

Fig. 11.1 Representation of a beam with a tapered wedge profile

Table 11.1 Parameters of the mechanical system

Parameter	Beam	ABH	Piezoelectric patch
Length	$L = 700$ mm	$L_{\mathrm{ABH}} = 700$ mm	67.5 mm
Width (b)	14 mm	14 mm	14 mm
Thickness	$h_0 = 14$ mm	$m = 2,\ x_0 = 10$ mm	2 mm
Young modulus	$E_{\mathrm{b}} = 210$ GPa	$E_{\mathrm{ABH}} = 210$ GPa	66 GPa
Density	$\rho_{\mathrm{b}} = 7800\ \mathrm{kg\,m^{-3}}$	$\rho_{\mathrm{ABH}} = 7800\ \mathrm{kg\,m^{-3}}$	$7800\ \mathrm{kg\,m^{-3}}$
Damping	$\xi_{\mathrm{b}} = 0.05\%$	$\xi_{\mathrm{ABH}} = 5\%$	

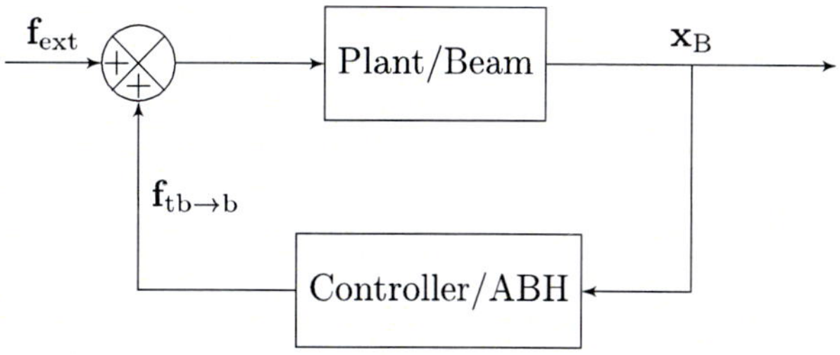

Fig. 11.2 Block diagram of the system

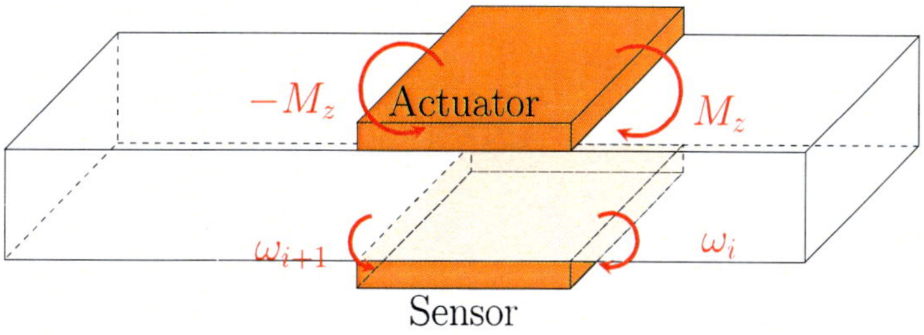

Fig. 11.3 Illustration of the piezoelectric patch on a portion of the beam

accuracy within the excitation frequency range [0 Hz, 2 kHz]. Moreover, the evaluation of (11.3b) can also be computed through a state-space formulation [13] for simplicity.

In the active control terminology, the dynamics of the plant (beam) has a feedback function with a controller (ABH). By measuring the displacement at the boundary ($\mathbf{x}_{\text{B}}$), one can actuate the forces ($\mathbf{f}_{\text{tb}\rightarrow\text{b}}$) to mimic the effect of the physical ABH. This process is illustrated in Fig. 11.2.

To reproduce exactly the ABH effect, all displacements of the boundary must be captured, and all forces need to be applied. This constitutes a limitation to the active control approach as no device can apply at the same point both independent forces and a torque. Therefore, in the following, the approach will measure the rotation at the boundary and apply a torque to reproduce the rotation effect of the ABH.

11.3 Analysis

The rotational effect of the ABH is achieved with a rectangular piezoelectric patch [14]. Figure 11.3 illustrates a collocated pair on a portion of a beam. Due to the deformation of the structure, the sensor patch will provide a current $\dot{q}_{\text{s}}$ from which the rotation ω can be retrieved. The signal is then processed by the digital unit and a voltage will be delivered to the actuator patch V_{a}. Finally, this voltage will create a moment at the boundaries and thus reproduce the rotational effect of the ABH.

However, a rectangular piezoelectric patch creates two moments: one at each of its boundaries. Based on the theory presented in the previous section, the ABH rotational effect should only be applied to the extremity of the beam. To remedy this issue, a first idea would be to create a very long patch which covers the entire length of the beam. The correct ABH torque could be applied at the free tip, and the additional torque would be applied to the clamped side and hence will not impact the result.

In practice, very long patches are not convenient. Another solution would be to consider multiple adjacent patches one beside the other to cover the length of the beam. The patch at the free tip will reproduce the ABH effect, while its neighbour will create a torque to compensate the additional torque created by the first patch, and so on. This approach is detailed next theoretically.

We consider a beam composed of N pairs of identical collocated piezoelectric patches, see Fig. 11.4. Each pair is adjacent in order to cover the entire length of the beam. In one pair, as shown in Fig. 11.3, one piezoelectric patch is actuating the structure, whereas the other is sensing its response. This sensor is assumed to be connected to a perfect current sensing device and is thus considered to be in short circuit ($V_{i,\text{s}} = 0$). The equations of motion of the system are

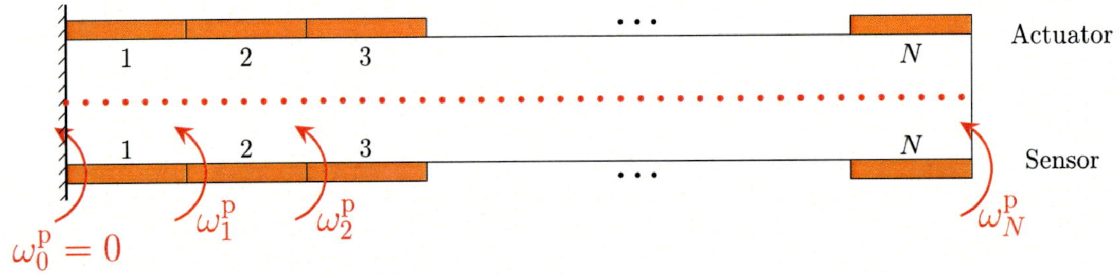

Fig. 11.4 Beam covered with N pairs of collocated piezoelectric patches

$$\begin{cases} \mathbf{M}\ddot{\mathbf{x}} + \mathbf{C}\dot{\mathbf{x}} + \mathbf{K}_{\mathrm{sc}}\mathbf{x} + \sum_{i=1}^{N} (\boldsymbol{\Theta}_i C^{\varepsilon} V_{i,\mathrm{a}}) = \mathbf{F} \\ \boldsymbol{\Theta}_i^{\mathsf{T}}\mathbf{x} - (C^{\varepsilon})^{-1} q_{i,\mathrm{a}} = V_{i,\mathrm{a}}, & \forall i \in [\![1, N]\!] \\ C^{\varepsilon}\boldsymbol{\Theta}_i^{\mathsf{T}}\mathbf{x} = q_{i,\mathrm{s}}, & \forall i \in [\![1, N]\!]. \end{cases} \tag{11.4}$$

The scalar C^{ε} represents the capacitance at constant strain of the piezoelectric patches. The vector $\mathbf{x}$ contains the displacements of all degrees of freedom (of size $3 \times N_{\mathrm{nodes}}$),

$$\mathbf{x} = \left[u_1, v_1, \omega_1, \cdots, u_j, v_j, \omega_j, \cdots, u_k, v_k, \omega_k, \cdots, u_{N_{\mathrm{nodes}}}, v_{N_{\mathrm{nodes}}}, \omega_{N_{\mathrm{nodes}}}\right]^{\mathsf{T}}, \tag{11.5}$$

where j and k are two arbitrary nodes. The matrix $\boldsymbol{\Theta}_i$ is the i-th piezoelectric patch coupling vector. Assuming that the boundaries of the i-th piezoelectric patch are connected to the nodes j and k, we have (following the same layout as Equation (11.5))

$$\boldsymbol{\Theta} = \theta\,[0, 0, 0, \cdots, 0, 0, -1, \cdots, 0, 0, 1, \cdots, 0, 0, 0]^{\mathsf{T}}, \tag{11.6}$$

with θ a piezoelectric constant equal for all patches. The matrix $\mathbf{K}_{\mathrm{sc}}$ is the stiffness matrix of the short-circuited system. The finite element method [15] is employed to compute all the quantities (dynamics of the piezoelectric patches accounted for).

In the following, ω_i^{p} denotes the rotation of the right boundary of the i-th piezoelectric patch as depicted in Fig. 11.4.

The voltage applied to the structure $V_{i,\mathrm{a}}$ is specified by the impedance Z_i implemented in the digital unit. In the Laplace domain, we have

$$\begin{aligned} V_{i,\mathrm{a}} &= Z_i(s) Q_{i,\mathrm{s}} \\ &= Z_i(s)\, C^{\varepsilon}\Theta_i^{\mathsf{T}}\mathbf{X} \\ &= Z_i(s)\, C^{\varepsilon}\theta\left(\Omega_i^{\mathrm{p}} - \Omega_{i-1}^{\mathrm{p}}\right), \end{aligned} \tag{11.7}$$

where Q, $\mathbf{X}$, and Ω^{p} are the Laplace transform of q, $\mathbf{X}$, and ω^{p}, respectively. Substituting this equation in the dynamics of the beam (first equation of (11.4)) and equating the result to the dynamics of the beam with ABH (see Equation (11.2a)), we have the following equations to satisfy:

$$\begin{aligned} (C^{\varepsilon}\theta)^2 Z_N \left(\Omega_N^{\mathrm{p}} - \Omega_{N-1}^{\mathrm{p}}\right) &= -Z_{\mathrm{ABH}}\Omega_N^{\mathrm{p}} \\ (C^{\varepsilon}\theta)^2 Z_i \left(\Omega_i^{\mathrm{p}} - \Omega_{i-1}^{\mathrm{p}}\right) &= (C^{\varepsilon}\theta)^2 Z_{i+1}\left(\Omega_{i+1}^{\mathrm{p}} - \Omega_i^{\mathrm{p}}\right) \forall i \in [\![1, N-1]\!], \end{aligned} \tag{11.8}$$

with $\Omega_0^{\mathrm{p}} = 0$ and Z_{ABH} the rotational part of $\mathbf{Z}_{\mathrm{ABH}}$ defined in Equation (11.3b). The first equation translates the ABH effect to reproduce. The second equation corresponds to torques created by two adjacent patches which need to be cancelled out. The solution of the previous system is

$$Z_i = \frac{-Z_{\mathrm{ABH}}}{(C^{\varepsilon}\theta)^2}\left(\frac{\Omega_N^{\mathrm{p}}}{\Omega_i^{\mathrm{p}} - \Omega_{i-1}^{p}}\right) \forall i \in [\![1, N]\!]. \tag{11.9}$$

Summing the measured current (see the third part of Equation (11.4)) gives

$$\Omega_N^{\mathrm{p}} = \frac{1}{(C^{\varepsilon}\theta)} \sum_{i=1}^{N} Q_{i,\mathrm{s}}. \tag{11.10}$$

Finally, the impedance of (11.9) is equal to

$$Z_i = \frac{-Z_{\mathrm{ABH}}}{(C^{\varepsilon}\theta)^2 \, q_{i,\mathrm{s}}} \sum_{i=1}^{N} q_{i,\mathrm{s}} \; \forall i \in [\![1, N]\!]. \tag{11.11}$$

In the end, the voltage that needs to be applied to the piezoelectric patch is equal to

$$\begin{aligned} V_{i,\mathrm{a}} &= Z_i q_{i,\mathrm{s}} \\ &= \frac{-Z_{\mathrm{ABH}}}{(C^{\varepsilon}\theta)^2} \sum_{i=1}^{N} q_{i,\mathrm{s}}. \end{aligned} \tag{11.12}$$

Therefore, in practice, the sum of the currents is measured (which can be achieved by connecting the sensors in parallel), the function $\frac{-Z_{\mathrm{ABH}}}{(C^{\varepsilon}\theta)^2}$ is implemented in the digital unit, and the actuation voltage, being identical for every actuator, is applied in parallel to them.

The proposed approach is now validated numerically. Ten pairs of piezoelectric patches are attached to the structure. The geometry and material are provided in Table 11.1. The capacitance C^{ε} is equal to 25 nF. The vector $\boldsymbol{\Theta}$ is obtained with the finite element method [15]. A transversal excitation is applied at 65% of the beam, and the transversal displacement of the tip is observed.

In practice, gaps (of length 2.5 mm) exist between the patches, and thus the torques at boundaries do not cancel each other out perfectly: a residual moment exists. To study the influence of the VABH, four simulations are proposed: the beam itself, the beam with the full ABH effect, the beam with the VABH when no gap exists between the piezoelectric patches (ideal VABH), and finally the beam with VABH with gaps between the patches (realistic VABH). The discretization in the finite element model is such that nodes exist within the gaps. The results are presented in Fig. 11.5. When the ABH effect is taken into account (full one or digital), vibration reduction is obtained. Only the first mode at 24.5 Hz (below the cut-on frequency) is not well attenuated. Similar trends exist between the full ABH and the VABH: vibration mitigation is obtained after the cut-on frequency, and the resonant frequencies of the system are shifted to low frequencies. The VABH with realistic gaps provides very close results to the VABH without gap.

Overall, these results give very good confidence in achieving vibration reduction with VABH.

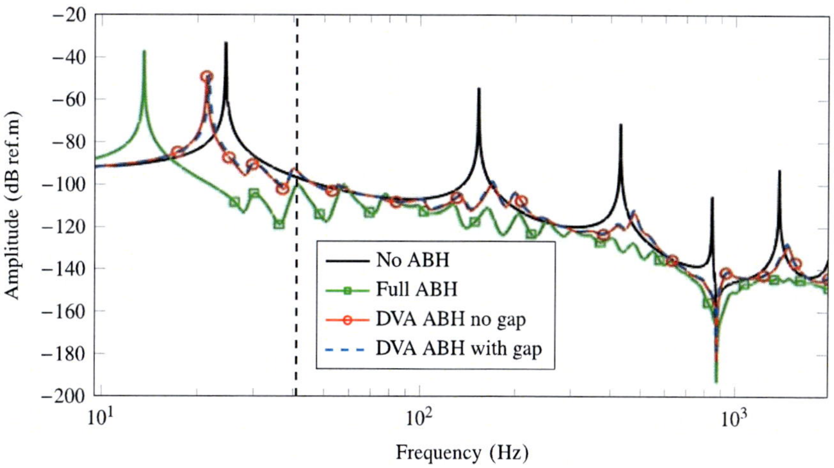

Fig. 11.5 Frequency forced response with multiple piezoelectric patches. The vertical dashed black line represents the cut-on frequency

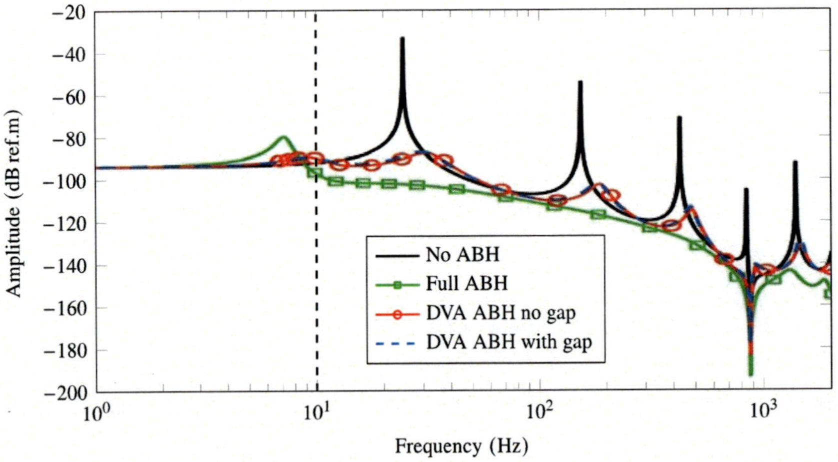

Fig. 11.6 Frequency forced response with a longer ABH

To show the versatility of our approach, we now create a VABH with a length of 1400 mm (and thus a cut-on frequency of 10 Hz) and a modal damping of 50%. The results are shown in Fig. 11.6. With the longer tapered wedge beam and the higher modal damping, all peaks are very well attenuated.

11.4 Conclusion

This chapter presents the combination of the acoustic black hole effect with digital controllers to create for the first time a VABH. Piezoelectric patches were employed to realize this device. As the ABH consists in applying appropriate boundary conditions at the tip of the beam, multiple patches were considered to achieve this properly. The electromechanical equations of motion were presented and the mechanical impedance that needs to be implemented in the controller was derived.

The concept was then tested numerically, and excellent performance was observed. Besides, an ABH not realizable in practice was implemented successfully in the digital controller. The perspective of this work would be the experimental validation of these results.

Acknowledgments This research is supported by a grant from the Belgian National Science Foundation (FRS-FNRS PDR T.0124.21), which is gratefully acknowledged.

References

1. Mironov, M.: Propagation of a flexural wave in a plate whose thickness decreases smoothly to zero in a finite interval. Sov. Phys. Acoust. **34**, 318–319 (1988)
2. Pelat, A., Gautier, F., Conlon, S.C., Semperlotti, F.: The acoustic black hole: a review of theory and applications. J. Sound Vib. **476**, 115316 (2020)
3. Krylov, V.V., Tilman, F.J.B.S.: Acoustic 'black holes' for flexural waves as effective vibration dampers. J. Sound Vib. **274**(3), 605–619 (2004)
4. Zhang, L., Kerschen, G., Cheng, L.: Electromechanical coupling and energy conversion in a PZT-coated acoustic black hole beam. Int. J. Appl. Mech. **12**(08), 2050095 (2020). World Scientific Publishing Co
5. Denis, V., Pelat, A., Touzé, C., Gautier, F.: Improvement of the acoustic black hole effect by using energy transfer due to geometric nonlinearity. Int. J. Non-Linear Mech. **94**, 134–145 (2017)
6. Cheer, J., Hook, K., Daley, S.: Active feedforward control of flexural waves in an Acoustic Black Hole terminated beam. Smart Mater. Struct. **30**(3), 035003 (2021). IOP Publishing
7. Forward, R.L.: Electronic damping of vibrations in optical structures. Appl. Opt. **18**(5), 690–697 (1979). Optica Publishing Group
8. Hagood, N.W., von Flotow, A.: Damping of structural vibrations with piezoelectric materials and passive electrical networks. J. Sound Vib. **146**(2), 243–268 (1991)
9. Fleming, A.J., Behrens, S., Moheimani, S.O.R.: Synthetic impedance for implementation of piezoelectric shunt-damping circuits. Electron. Lett. **36**(18), 1525–1526 (2000). IET Digital Library

10. Raze, G., Jadoul, A., Guichaux, S., Broun, V., Kerschen, G.: A digital nonlinear piezoelectric tuned vibration absorber. Smart Mater. Struct. **29**(1), 015007 (2019). IOP Publishing
11. Aklouche, O., Pelat, A., Maugeais, S., Gautier, F.: Scattering of flexural waves by a pit of quadratic profile inserted in an infinite thin plate. J. Sound Vib. **375**, 38–52 (2016)
12. Craig, R., Bampton, M.: Coupling of substructures for dynamic analyses. AIAA J. **6**(7), 1313–1319 (1968)
13. Preumont, A.: State space approach. In: Preumont, A. (ed.) Vibration Control of Active Structures: An Introduction, 3rd edn. Solid Mechanics and Its Applications, pp. 187–213. Springer, Dordrecht (2011)
14. Preumont, A.: Piezoelectric beam, plate and truss. In: Preumont, A. (ed.) Vibration Control of Active Structures: An Introduction, 3rd edn. Solid Mechanics and Its Applications, pp. 61–101. Springer, Dordrecht (2011)
15. Thomas, O., Deü, J.-F., Ducarne, J.: Vibrations of an elastic structure with shunted piezoelectric patches: efficient finite element formulation and electromechanical coupling coefficients. Int. J. Numer. Methods Eng. **80**(2), 235–268 (2009)

Chapter 12
A Portable Fixed Base Support for Modal Survey Tests

Peter A. Kerrian, Kevin Napolitano, and Gregory Less

Abstract Fixed base (FB) correction methods (FBCMs) have been increasingly used to transform flexible or dynamically active boundary conditions into fixed boundaries in modal tests. This chapter documents the development of a portable test setup to conduct an FB modal survey test that can be integrated into other environmental testing configurations to accelerate testing schedules. The test article is attached to a T-slot table via an adapter plate that is bolted at multiple locations via T-slot bolts. During testing, inflatable airbags are used to raise the T-slot table off the supports, thus creating a soft boundary condition. Using 11 electrodynamic modal shakers, nine constraint shapes were measured and used to remove the dynamics of the T-slot table via the FBCM. Overall, there was excellent agreement between the FB target modes and the extracted FB modes. With this deployable setup, FB modal tests can easily be integrated into existing testing schedules, providing the ability to correlate FB models of flight hardware.

Keywords Modal testing · Vibrations · Base-shake · Fixed base · Constraint shapes · Boundary condition correction

Nomenclature

ATA	ATA Engineering Inc.
CMIF	Complex mode indicator function
CS	Constraint shape
DOF	Degree of freedom
DP	Drive point
FB	Fixed base
FBCM	Fixed base correction method
FEM	Finite element model
FRF	Frequency response function
IMAT™	Interface between MATLAB Analysis and Test
MSS	Millennium Space Systems
MST	Modal survey test
PODIuM	Portable Dynamically Fixed Mass
PSMIF	Power spectral mode indicator function
RB	Rigid body
SMURF	Structural modification using frequency response functions
SVD	Singular value decomposition
TA	Test article

12.1 Introduction

Modal survey tests (MSTs) are performed on flight hardware to help ground finite element models (FEMs) to measured test data. Doing so increases confidence in the results from subsequent analyses, such as coupled loads analysis. Fixed base (FB)

P. A. Kerrian (✉) · K. Napolitano
ATA Engineering, Inc., San Diego, CA, USA

G. Less
Millennium Space Systems, El Segundo, CA, USA

C. Walber et al. (eds.), *Sensors & Instrumentation and Aircraft/Aerospace Testing Techniques, Volume 8*, Conference Proceedings of the Society for Experimental Mechanics Series, https://doi.org/10.1007/978-3-031-34938-6_12

mode shapes are the desired boundary condition for a correlation effort for two reasons. First, Hurty/Craig-Bampton analysis models of test articles (TAs) used in coupled loads analysis are fixed at their base interfaces. Second, having a fixed interface eliminates the need to spend time during the correlation updating the boundary conditions. Unfortunately, conducting an FB MST can be very costly to both the program schedule and budget because of the challenges associated with creating a support structure stiff enough to behave as a fixed boundary.

The FB correction method (FBCM) [1–10] was developed to transform flexible or dynamically active boundary conditions into fixed boundaries by using acceleration data or constraint shapes (CSs) as references when calculating frequency response functions (FRFs). By mathematically removing the compliance of the TA interface from the FRFs, FB modes can be extracted and used for subsequent model correlation. Recently, ATA Engineering, Inc., (ATA) developed the Portable Dynamically Fixed Mass (PODIuM) as a portable FBCM test setup to enable high-quality FB MSTs to be performed anywhere. The PODIuM consists of a T-slot table on top of a soft-suspension system to which a TA can be attached via an adapter plate. The advantage of this setup is that it eliminates the need to either ship the TA to a separate testing facility or invest a significant amount of resources developing a separate test fixture.

The primary focus of this chapter is to document the successful deployment of the PODIuM for an MST on the Millennium Space Systems (MSS) Aquila™ bus structure. The MSS Aquila is designed to take Class C/B payloads into orbit. MSS contracted ATA to perform the MST and then perform a model correlation utilizing the FB modes extracted using the FBCM. During the test, nine shakers were used to remove the six rigid body (RB) and first three flexible modes of the T-slot table. Subsequently, the FB modes were used to correlate the MSS Aquila FEM.

12.2 Test Overview

The MSS Aquila consists of a large rectangular bus structure, large solar panels connected at the top of the bus structure on two sides, two triangular antenna booms mounted to opposite corners of the top, and an antenna dish mounting structure connected to the top. For the test configuration, the bus structure was connected to an adapter ring, which was bolted to a second adapter ring to elevate the vehicle to allow for sufficient clearance for the solar panels in the test configuration. The second adapter ring was mounted to a 1 in. thick adapter plate that was mounted to the T-slot table with rings of T-slot bolts on the inside and outside of the footprint of the secondary adapter ring. During testing, the T-slot table was raised off the rigid ground supports using a soft-suspension system that consisted of four air bag isolators. The entire boundary configuration was added to the FEM of the MSS Aquila, as shown in Fig. 12.1.

A traditional pretest analysis was performed to identify sensor placements and create a back-expansion matrix for the MST. In total, 231 acceleration degrees of freedom (DOFs) at 153 separate locations were measured simultaneously using multiple LAN-XI 12-channel 3053 modules. Nine shakers were positioned around the T-slot table for use as references in the FBCM, as shown in Fig. 12.2. Six vertical shakers were positioned to remove the first three out-of-plane RB modes (RB Z, RX, and RY) and the first three flexible T-slot table modes (first bending, first torsion, and second torsion); three lateral shakers were positioned to remove the first three in-plane RB modes (RB X, Y, and RZ).

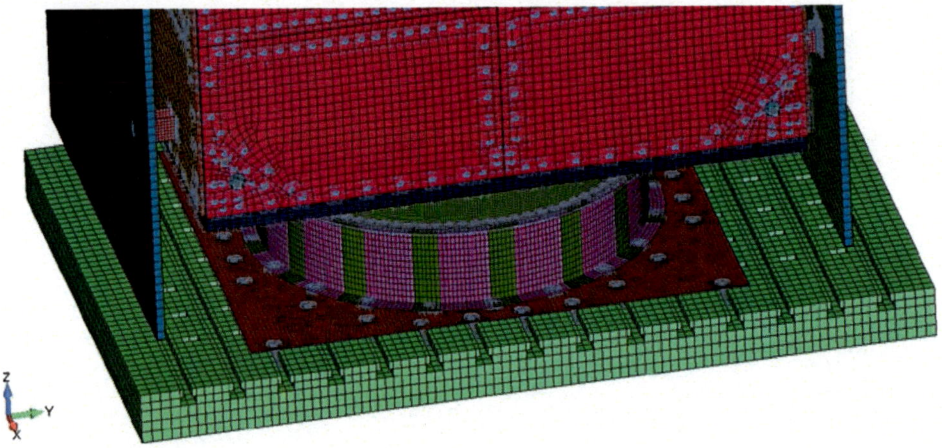

Fig. 12.1 FEM visualization of the base support structure for the MSS Aquila™ MST

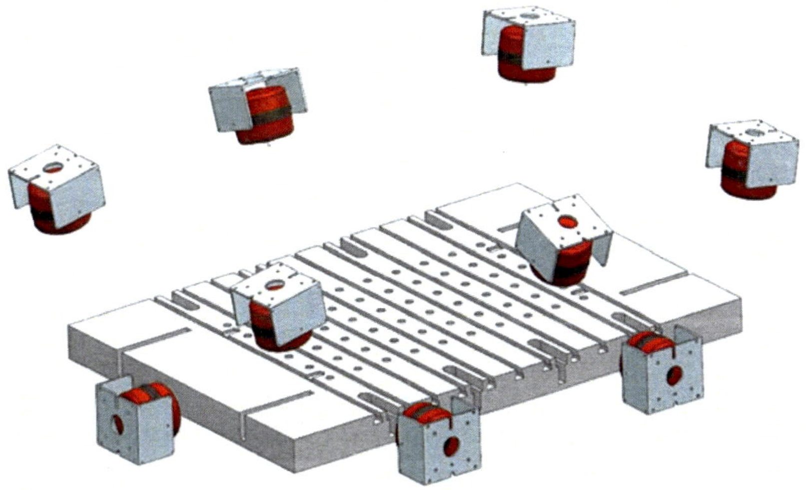

Fig. 12.2 Depiction of the six vertical and three lateral shakers attached to the T-slot table

Two additional shakers were attached to the TA for excitation of the TA modes. The shaker controller used was an EMX-1434 system driven by a custom MATLAB/IMAT™ software application. All 11 shakers were excited with continuous random vibration at three different levels. Impact data were collected at 25 locations on the MSS Aquila and at 9 locations on the T-slot table.

12.3 Shaker Data Analysis

The theory of the FBCM has been previously explained in multiple publications [3, 5, 8, 9]. This section focuses on the implementation of the method using the raw time histories collected from the shaker data. All data processing was performed using ATA's IMAT™ suite of MATLAB-based software applications. Upon completion of each run, FRFs were computed for the free test configuration to evaluate data quality. A complex mode indicator function (CMIF) was computed, shown in Fig. 12.3, to evaluate the overall modal behavior of the test system. In anticipation of utilizing the FBCM, a useful secondary data quality check is examining the antiresonances of the CMIF of the FRF matrix partitioned down to the DOF associated with the drive points (DPs) on the T-slot table, as shown in Fig. 12.4. If structural modification using FRFs (SMURF) [1] was used to calculate FB FRFs directly from the DP accelerations, this subset of the FRF matrix directly shows the matrix inversion that would be performed. The antiresonances oftentimes become the resonances in the FB FRFs. Figure 12.4 provides a good example of the type of data quality needed for the calculation of clean FB FRFs.

Utilizing SMURF to calculate FB modes from the DP forces and accelerations only fixes those individual DOFs. Alternatively, CS can be used as references in the FRF calculation to remove additional dynamics of the T-slot table. Because the T-slot table is isolated from the floor with a soft-suspension system, the six traditional RB modes were used as the first six CSs. Additional CSs are calculated from the measured time histories with a singular value decomposition (SVD). For the MSS Aquila MST, three CSs were extracted from the time histories; these CSs are considered the first three flexible modes of the T-slot table. The FBCM requires one unique reference (excitation source) per CS. Table 12.1 shows the nine CSs extracted from the DOFs on the T-slot table. These CSs allowed for calculation of FB FRFs up to 100 Hz. To extend the analysis to higher frequencies, additional shaker references can be used to calculate higher-order CSs associated with the next flexible modes of the T-slot table. Rather than utilizing SMURF with CS FRFs, the following procedure was used to directly calculate the FB FRFs from the time-history data [7]:

1. Calculate the six RB modes of the T-slot table based on the test display model geometry.
2. Estimate RB time histories based on the response DOF time histories on the T-slot table.
3. Subtract the RB time histories from the original time-history data.
4. Perform an SVD of the remaining signal to obtain three flexible CSs.
5. Calculate CS time histories.
6. Calculate FB FRF utilizing the H_{SVD} method [11] with the following settings:

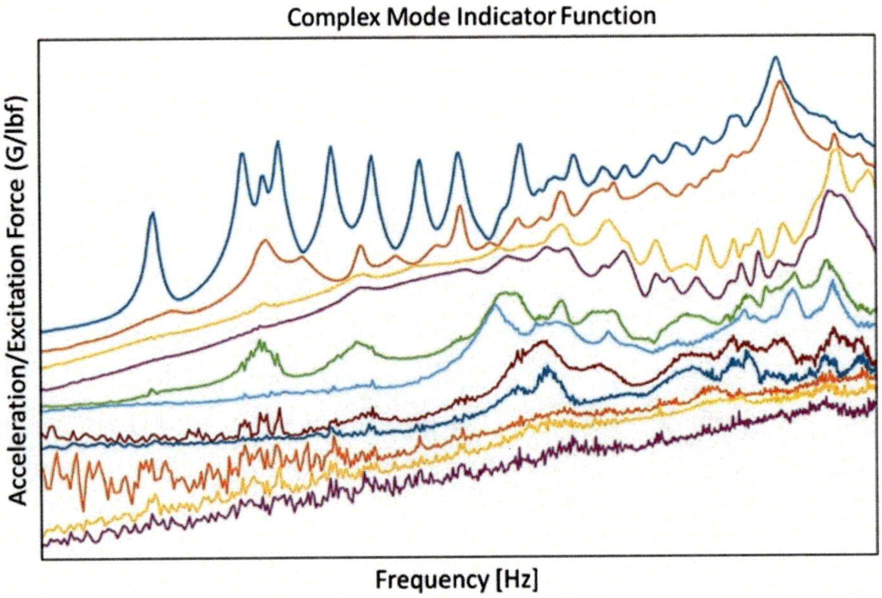

Fig. 12.3 Overall CMIF of free raw test data

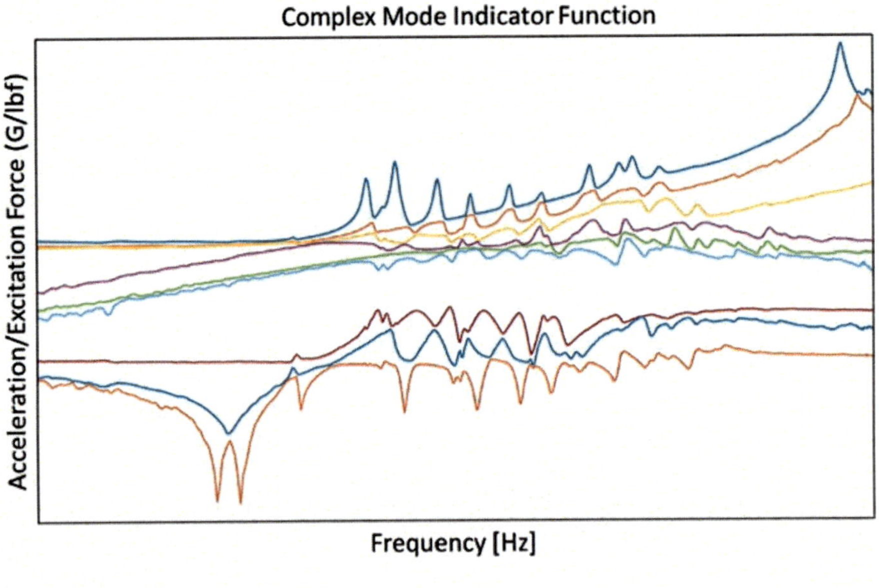

Fig. 12.4 Overall CMIF of FRF matrix inversion

- *References*: Two TA excitation forces and nine CS virtual time histories
- *Basis vectors*: Two TA excitation forces and nine T-slot table excitation forces
- *Responses*: Two TA DP accelerations, all response DOFs, and nine T-slot table excitation forces and accelerations

7. Calculate FB modes utilizing standard parameter estimation software.

Figure 12.5 shows a CMIF of the FB FRFs for the MSS Aquila. As previously mentioned, the antiresonances of the CMIF in Fig. 12.4 correspond to the resonances of the CMIF in Fig. 12.5. Mode shapes were extracted from the FB FRFs using ATA's IMAT modal parameter application OPoly™. Table 12.2 shows the first four extracted modes of the MSS Aquila. Visualization of the mode shapes indicates that the dynamics of the T-slot table have been removed. Although not included in this chapter, standard linearity checks can be made utilizing the FB FRFs for different excitation force levels.

Table 12.1 Summary of T-slot table CSs used for FB FRF calculation

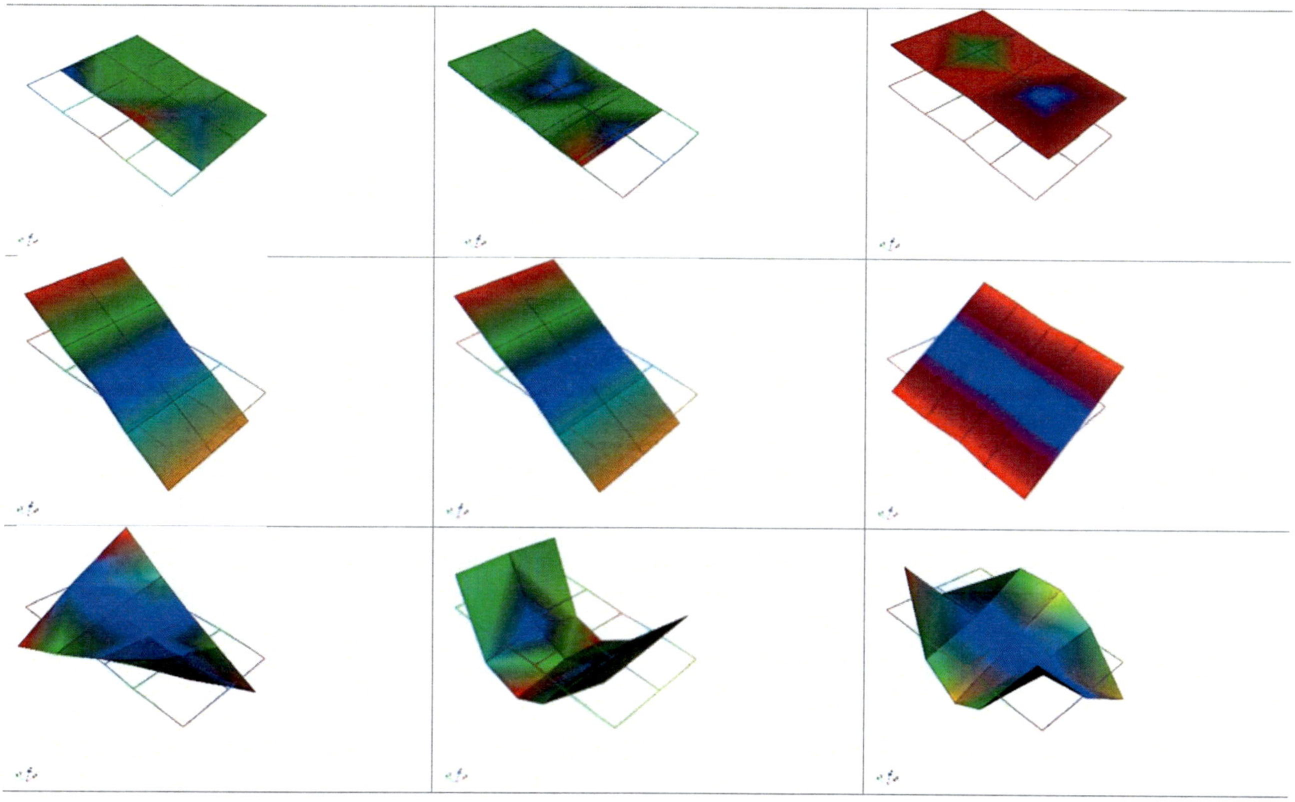

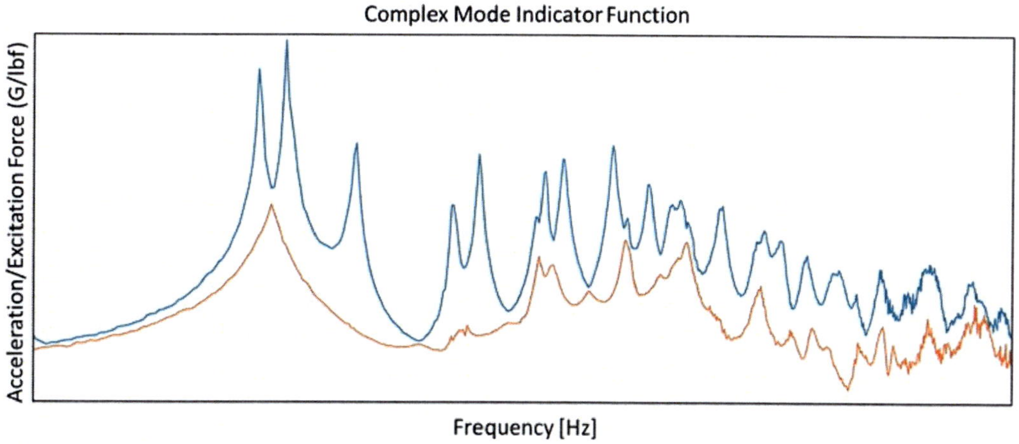

Fig. 12.5 CMIF of FB-corrected FRF matrix

12.4 Impact Data Analysis

For the MSS Aquila MST, the majority of the target modes were extracted from multipoint random shaker excitation, as discussed in the previous section. However, as with most MSTs, impact data were needed to supplement the shaker data to extract higher-order component modes. To illustrate the process of utilizing impacts with the FBCM, data from impacts on one antenna boom of the MSS Aquila are presented with the objective of extracting the second-order bending and torsion modes, which were not well excited with the shaker data. Impacts were made in two orthogonal directions on the boom as well as on the T-slot table in the same locations as the shaker references. The following process was used to compute FB FRFs and extract the component modes. All local component FRFs could be combined into a single matrix for processing.

Table 12.2 First four mode shapes of the MSS Aquila

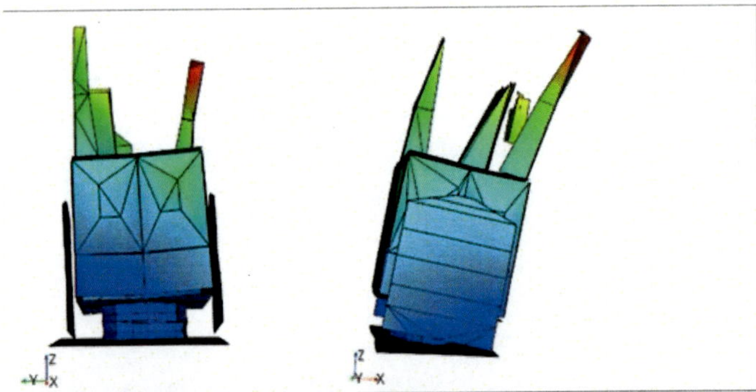	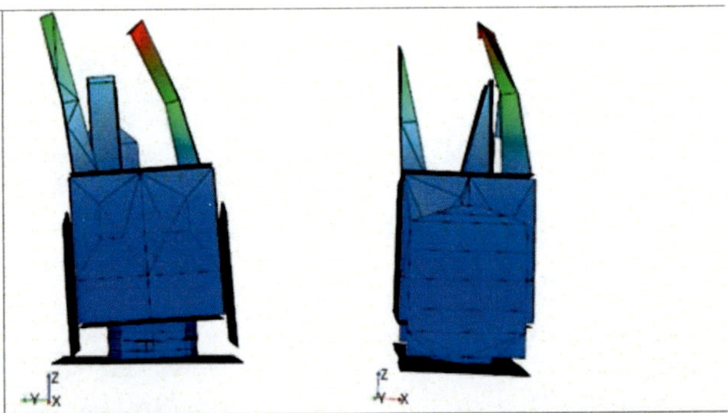
Global rocking X	Global rocking Y
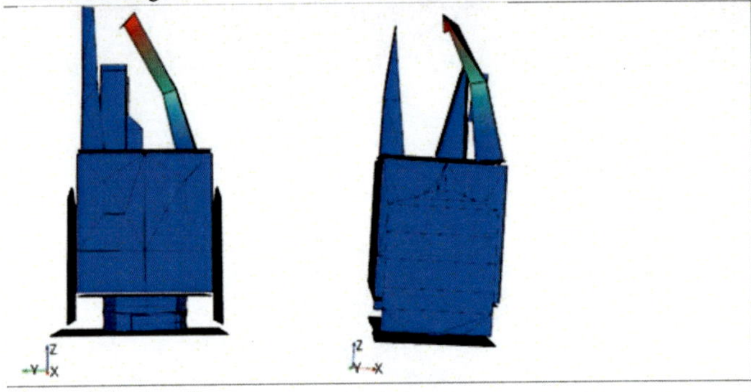	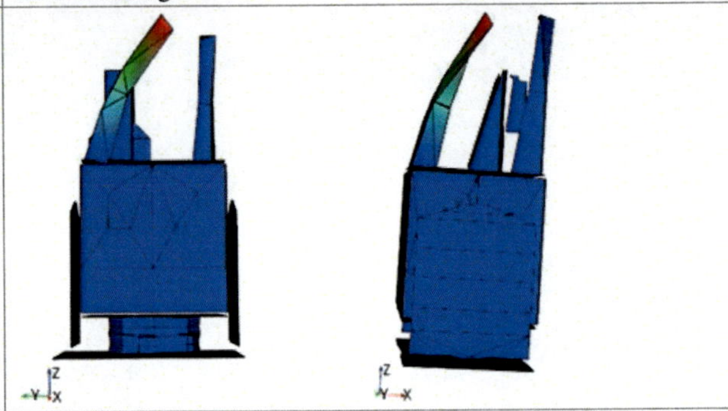
Boom 1 first bending	Boom 2 first bending

1. Calculate traditional acceleration/force FRFs for each impact location.
2. Combine all impact data into a single FRF matrix.
3. Use a partial inversion of the FRF matrix to calculate FB FRFs by moving the DP forces to responses and moving either a) the T-slot table DP accelerations or b) the CS DOFs to the references.
4. Extract modes for each component using standard modal parameter software using only the DOF on the component for the curve-fitting process for cleaner pole estimates.

Table 12.3 shows the primary modes of one of the MSS Aquila antenna booms. The torsion and second bending modes were cleanly extracted from the impact data. Figure 12.6 shows the power spectral mode indicator function (PSMIF) for only the DOFs on the boom for both the uncorrected and the FB-corrected FRFs. As indicated by the PSMIF, shifts in the frequency of the first bending modes show the influence of the T-slot table dynamics. However, the higher-order torsion and second bending modes were not affected by the T-slot table dynamics and could have been extracted directly from the measured FRFs. This is typical of most MSTs, where higher-frequency (and low-effective-mass) local component modes are not affected by TA boundary condition.

12.5 Posttest Model Correlation

The final set of mode shapes for the MSS Aquila MST was compiled from a combination of shaker and impact data, as described in the previous sections. In total, 55 FB modes were extracted. Table 12.4 shows the cross-orthogonality matrix and summary table comparison between the FEM and FB test shapes at the conclusion of the test. In general, there is decent agreement with respect to shape orthogonality, but there is a significant frequency discrepancy, which indicates that a significant amount of stiffness was missing from the model.

The model updating process consisted primarily of introducing stiffness to the bolted connections throughout the model and updating simplified meshes of component joints to better represent the as-built component. Adjustments were also made to spring stiffnesses and mass moments of inertia to certain lumped-mass elements. Table 12.5 shows that there was significant improvement to the cross-orthogonality matrix and summary comparison table between the updated FEM and the FB test shapes. Typical correlation standards are for diagonal orthogonality values to be above 90%, for off-diagonal values

Table 12.3 Primary modes of antenna boom

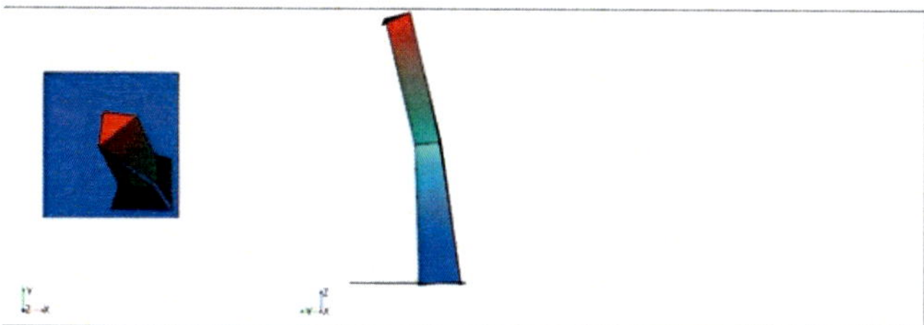

First boom bending

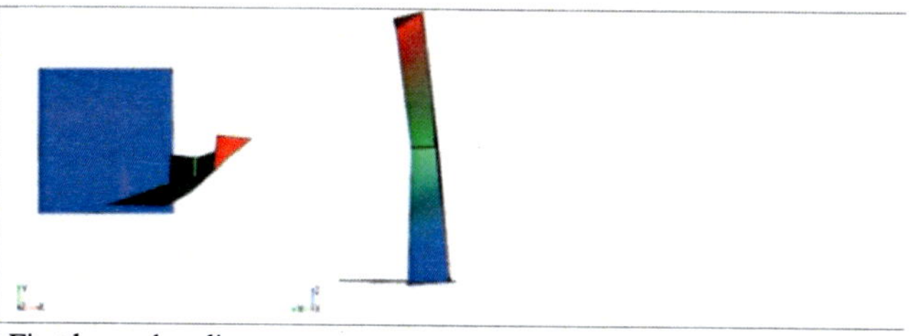

First boom bending

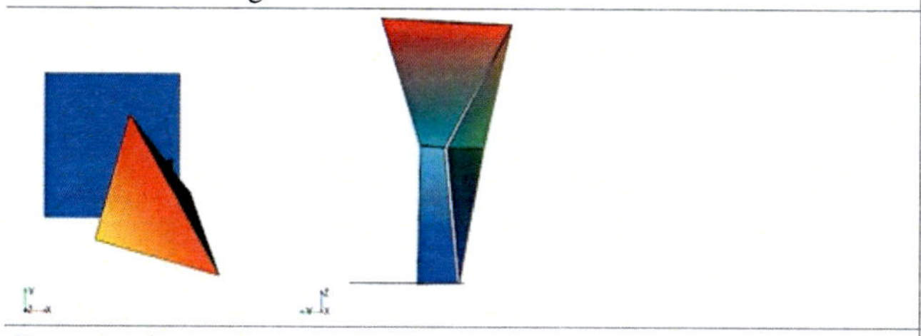

Boom torsion

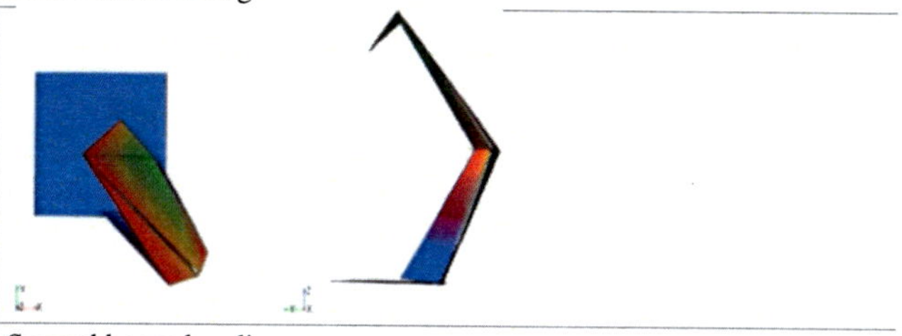

Second boom bending

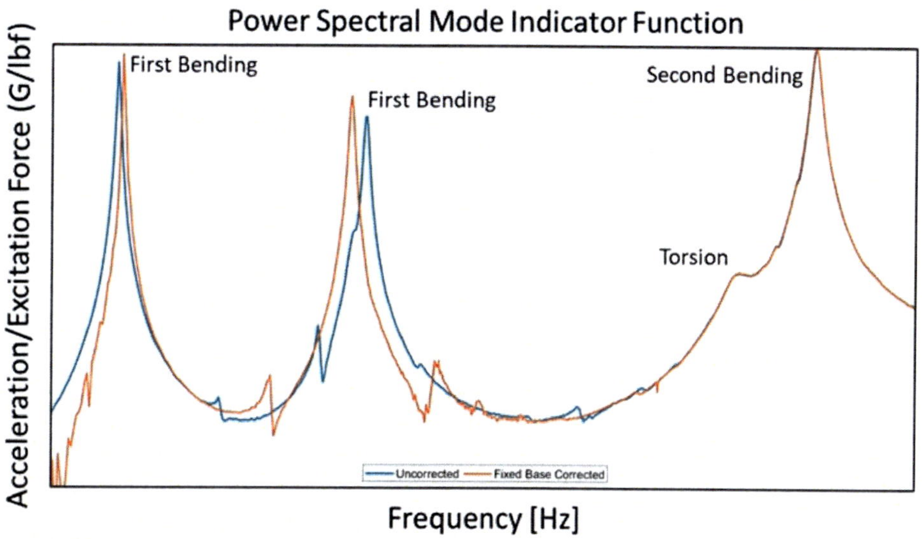

Fig. 12.6 PSMIF of impacts before and after FB correction

Table 12.4 Posttest cross orthogonality

FEM/Test Cross Orthogonality Table

FEM Shapes

Otg	Test S	1	2	3	4	5	6	7	8	9	10	11	12	13	14	15	16	17	18	CRSS (3%)
100	1	97									22									97
95	2		98																	98
90	3		21	98																98
85	4				25	96														99
80	5				93	26														97
75	6							67	63											67
70	7						97													97
65	8							87	44											87
60	9								35	91										91
55	10							24		25	47	91								92
50	11										68	28	49	52	34	35				72
45	12											20	77		43	37				81
40	13													35		69		25		69
35	14										28			42	22	48		50		52
30	15																		53	53
3%	CRSS	97	98	98	93	96	97	87	63	91	68	91	79	52	53	79	81	73	79	

Legend: 90 - 100; 50 - 89

Test Mode No.	FEM Mode No.	Freq Pct Diff	Cross Ortho	CRSS XOrtho 3%
1	1	-16.9	97	97
2	2	-17.2	98	98
3	3	-21.2	98	98
4	5	-22.8	96	99
5	4	-28.9	93	97
6	7	-19.4	67	67
7	6	-26.5	97	97
8	7	-24.2	87	87
9	9	-22.7	91	91
10	11	-17.1	91	92
11	10	-23.3	68	72
12	12	-22.1	77	81
13	15	-15.2	69	69
14	17	-11.1	50	52
15	18	-7.6	53	53

Table 12.5 Post-model-correlation cross orthogonality

FEM/Test Cross Orthogonality Table

3%	90 - 100
CRSS	50 - 89

Otg	Test	FEM 1	2	3	4	5	6	7	8	9	10	11	12	13	14	15	CRSS
100	1	99															99
95	2		100														100
90	3			100													100
85	4				99												99
80	5					99											99
75	6						99		37								99
70	7							98	24								98
65	8						25		96								98
60	9									97		20					97
55	10									20	95						95
50	11											96					96
45	12												97				98
40	13													90	33		97
35	14													20	93	20	96
30	15															97	97
3%	CRSS	99	100	100	99	99	99	98	96	97	95	96	99	98	95	98	

Test Mode No.	FEM Mode No.	Freq Pct Diff	Cross Ortho	CRSS XOrtho 3%
1	1	0.5	99	99
2	2	-0.1	100	100
3	3	1.8	100	100
4	4	-2.7	99	99
5	5	-0.9	99	99
6	6	-1.1	99	99
7	7	1.7	98	98
8	8	0.0	96	98
9	9	0.8	97	97
10	10	2.8	95	95
11	11	1.2	96	96
12	12	0.2	97	98
13	13	1.9	90	97
14	14	0.8	93	96
15	15	0.4	97	97

to be below 10, and for frequencies to match within 5%. The results of the MSS Aquila FEM match all requirements for diagonal values, and frequencies match to within 3% for the first 15 modes of the structure. The remaining high off-diagonal terms in the matrix are associated with limitations on where sensors could be placed because the TA was flight hardware.

12.6 Summary

Overall, two primary conclusions can be drawn from the results of the MSS Aquila™ MST. First, ATA's PODIuM provided the portable FB support test setup, which resulted in a successful MST. Utilizing the PODIuM significantly reduced the schedule and cost compared to a traditional FB MST because it eliminated the need to either build a true FB boundary condition at MSS's facility or move the TA to a remote facility with an adequate boundary condition. Additionally, the correlation effort was focused entirely on the MSS Aquila FEM and not on the boundary condition – leading to improved model results.

References

1. Imregun, M., Robb, D.A., Ewins, D.J.: Structural Modification and Coupling Dynamic Analysis Using Measured FRF Data. In: Proceedings of the 5th International Modal Analysis Conference (1987)
2. Mayes, R.L., Bridgers, L.D.: Extracting Fixed Base Modal Models from Vibration Tests on Flexible Tables. In: Proceedings of the 27th International Modal Analysis Conference (2009)
3. Napolitano, K., Yoder, N.: Fixed Base FRF Using Boundary Measurements as References – Analytical Derivation. In: Proceedings of the 30th International Modal Analysis Conference (2012)
4. Mayes, R., Rohe, D., Blecke, J.: Extending the Frequency Band for Fixed Base Modal Analysis on a Vibration Slip Table. In: Proceedings of the 31th International Modal Analysis Conference (2013)
5. Napolitano, K., Yoder, N.: Extraction of Fixed-Base Modes of a Structure Mounted on a Shake Table. In; Proceedings of the 31th International Modal Analysis Conference (2013)
6. Staab, L., Winkel, J., Suárez, J., Jones, T., Napolitano, K.: Fixed Base Modal Testing Using the Mechanical Vibration Facility 3-Axis Base Shake System. In: Proceedings of the 34th International Modal Analysis Conference (2016)
7. Winkel, J., Akers, J., Suarez, V., Staab, L., Napolitano, K.: Modal Survey of the MPCV Orion European Service Module Structural Test Article Using a Multi-Axis Shake Table. In: Proceedings of the 37th International Modal Analysis Conference (2018)
8. Napolitano, K. Fixing Degrees of Freedom of an Aluminum Beam by Using Accelerometers as References. In: Proceedings of the 38th International Modal Analysis Conference (2019)
9. Kerrian, P., Napolitano, K.: Pretest Analysis for Modal Survey Tests Using Fixed Base Correction Method. In: Proceedings of the 38th International Modal Analysis Conference (2019)
10. Kerrian, P.: A Comparison of Different Boundary Condition Correction Methods. In: Proceedings of the 39th International Modal Analysis Conference (2020)
11. Napolitano, K.: Using Singular Value Decomposition to Estimate Frequency Response Functions. In: Proceedings of the 29th International Modal Analysis Conference (2011)

Chapter 13
Characterization of Nonlinear Joint Stiffness Using Dynamic and Static Experimental Methods

Benjamin L. Martins, Caleb R. Heitkamp, and Joseph M. Jaeckels

Abstract An experimental study was conducted to characterize the nonlinear rotational stiffness of a missile fin control actuation system testbed. Dynamic testing was used to measure the acceleration response of the fin to a sine-sweep input at various force levels. The modal frequency and damping of the rotation mode were tracked as a function of applied force to determine the converged control surface rotation mode. Frequency response functions were generated in order to estimate the dynamic stiffness of the control actuation system joint. Static freeplay and rigidity data were also collected using an applied quasi-static load and measuring the resultant displacements at multiple locations around the fin. The freeplay and rigidity data were used to characterize the degree of nonlinearity in the joint stiffness with respect to applied load.

Keywords Control actuation system · Rotational stiffness · Nonlinear vibration · Freeplay testing

Nomenclature

ATA	ATA Engineering, Inc.
B&K	Brüel & Kjær
CAS	Control actuation system
CSLT	Control surface linearity test
DAS	Data acquisition system
F&R	Freeplay and rigidity
FEM	Finite element model
FRF	Frequency response function
IMAT	Interface between MATLAB, analysis, test
LVDT	Linear variable differential transducer
MAC	Modal assurance criteria
UUT	Unit under test

13.1 Introduction

Over the course of a typical aerospace vehicle development cycle, aeroelastic stability of the vehicle must be verified. This verification has historically come in the form of conducting structural dynamics tests and then correlating a finite element model (FEM) of the vehicle to the test data. Once correlated, the FEM is used to perform flutter analysis at various points in the flight envelope. The degree to which the correlated FEM represents the structural dynamics of the vehicle is thus of critical importance [1].

In addition to global flexible modes, control surface rotation modes are also of interest to ensure aeroelastic stability and controllability of the vehicle. Standard practice is to perform a control surface linearity test (CSLT), where increasing levels of dynamic excitation are used to excite a control surface's rotation mode until a converged frequency is reached. During the model correlation process, the control surface joint stiffness is then adjusted until the rotation mode of the control surface matches the converged experimental frequency. While this approach is fundamentally sound, it is possible that the amount

B. L. Martins · C. R. Heitkamp (✉) · J. M. Jaeckels
ATA Engineering, Inc., San Diego, CA, USA
e-mail: caleb.heitkamp@ata-e.com

C. Walber et al. (eds.), *Sensors & Instrumentation and Aircraft/Aerospace Testing Techniques, Volume 8*, Conference Proceedings of the Society for Experimental Mechanics Series, https://doi.org/10.1007/978-3-031-34938-6_13

of dynamic excitation applied to the test article is insufficient to characterize the joint stiffness beyond the freeplay region of the joint.

In this chapter, a study was performed which characterizes the stiffness of a control actuation system (CAS) joint testbed representative of a joint commonly found on missile fins. Dynamic testing was done to characterize the rotational frequency of the CAS fin and the frequency response functions (FRF) used to estimate the joint stiffness. A static freeplay and rigidity (F&R) test was also performed to characterize the joint stiffness to high enough force levels to extend out of the freeplay and nonlinear region.

13.2 Test Article

In order to simulate a missile fin CAS, a solid aluminum fin and adaptor were machined and coupled with a gearbox and stepper motor. The fin was trapezoidal in shape with a tip chord length of 4.125′′, a root chord length of 10.5′′, and a span of 7.75′′. The adapter was offset 0.5′′ in the direction of the leading-edge. The aluminum adapter attached the fin to the gearbox shaft using four bolts at each component. The gearbox was of the worm gear type with a gear speed reducer ratio of 30:1. The closed loop NEMA 23 stepper motor had a holding torque of 17.7 in-lb. Both the gearbox and stepper motor were purchased off the shelf. The fin, gearbox, and motor assembly were attached to a strongback mounted on a t-slot table, as shown in the schematic in Fig. 13.1. The setup was intended to provide a very stiff boundary condition in order to isolate the rotational flexibility of the fin.

13.3 Experiments

Separate test configurations and equipment were used to measure the nonlinear rotational stiffness of the CAS interface joint using both dynamic and static testing methods. An overview of the dynamic and static test setups and results is provided in the following subsections.

13.3.1 Dynamic Testing

Dynamic testing consisted of attaching an electrodynamic shaker to the fin to excite the structure while measuring the response at various locations using eight accelerometers. The shaker was attached to the trailing-edge root of the fin with accelerometers distributed about the fin. Fig. 13.2 provides an image of the test setup used for dynamic testing. An MB Dynamics Modal 110 shaker was chosen to excite the UUT, and PCB 352A21 teardrop accelerometers were selected for measuring the out-of-plane response. The excitation and response locations were chosen so as to best characterize the rigid

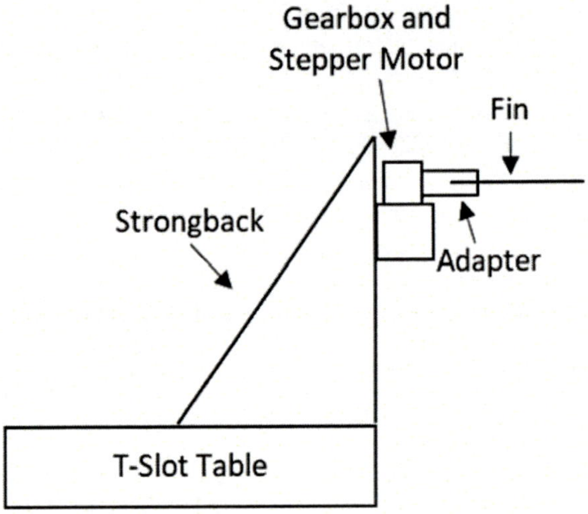

Fig. 13.1 Schematic of the unit under test (UUT) and overall test setup

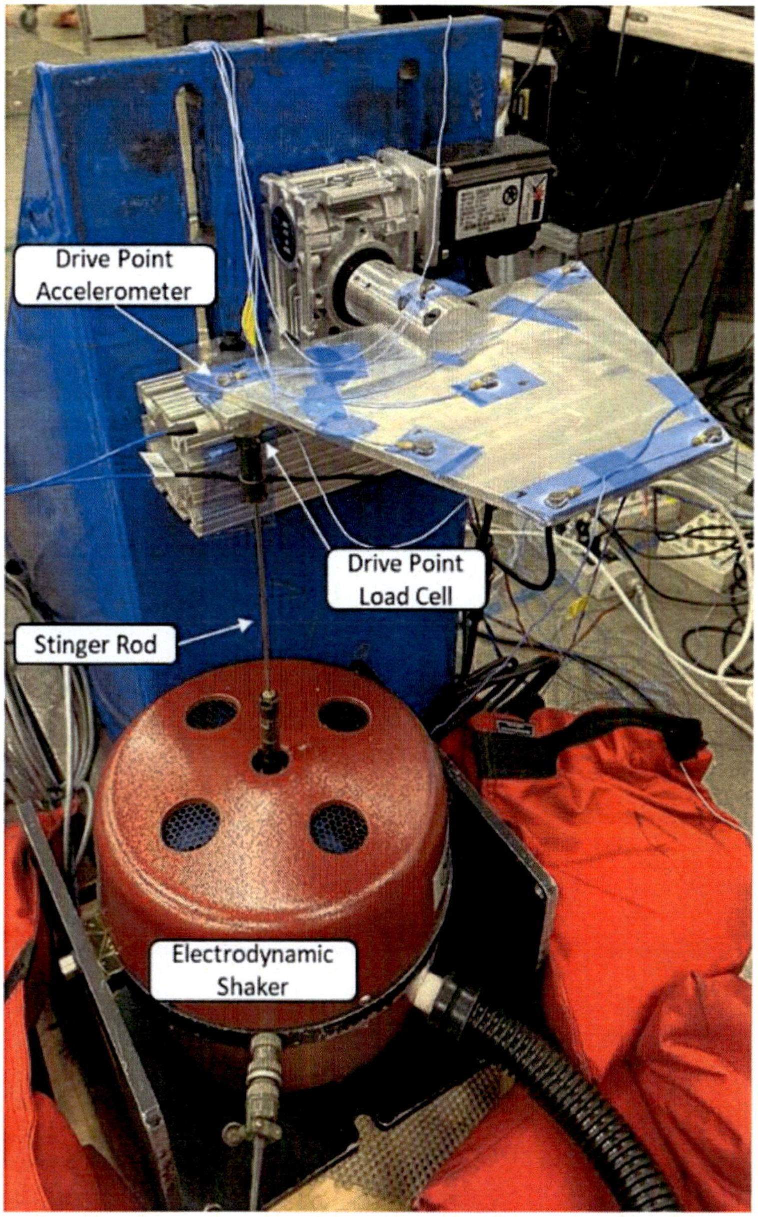

Fig. 13.2 Dynamic test setup

Table 13.1 Dynamic testing instrumentation used

Transducer	Model	Nominal sensitivity
Load cell	PCB 208C03	10 mV/lb
Accelerometer	PCB 352A21	10 mV/g
Shaker	MB Dynamics Modal 110	N/A

body rotation mode of the fin about the CAS hub. A load cell was placed at the shaker–fin interface to measure the applied dynamic force. A PCB 208C03 load cell with nominal sensitivity of 10 mV/lb. was used for the study. Table 13.1 lists the instrumentation used for the dynamic tests.

Sine sweeps were conducted over a range of up to 140 Hz to 5 Hz, while time-history data were recorded on a B&K LAN-XI data acquisition system (DAS). The time-history information of the load cell, accelerometers, and reference voltage was used to compute the FRF of the fin at various excitation force levels. The nonlinearity of the CAS interface joint proved problematic when computing the FRF via the typical "H_1" response acceleration over reference force FRF. It is thought that the out-of-band modes coupling with the modes over the excited frequency range caused poor coherence and, in turn,

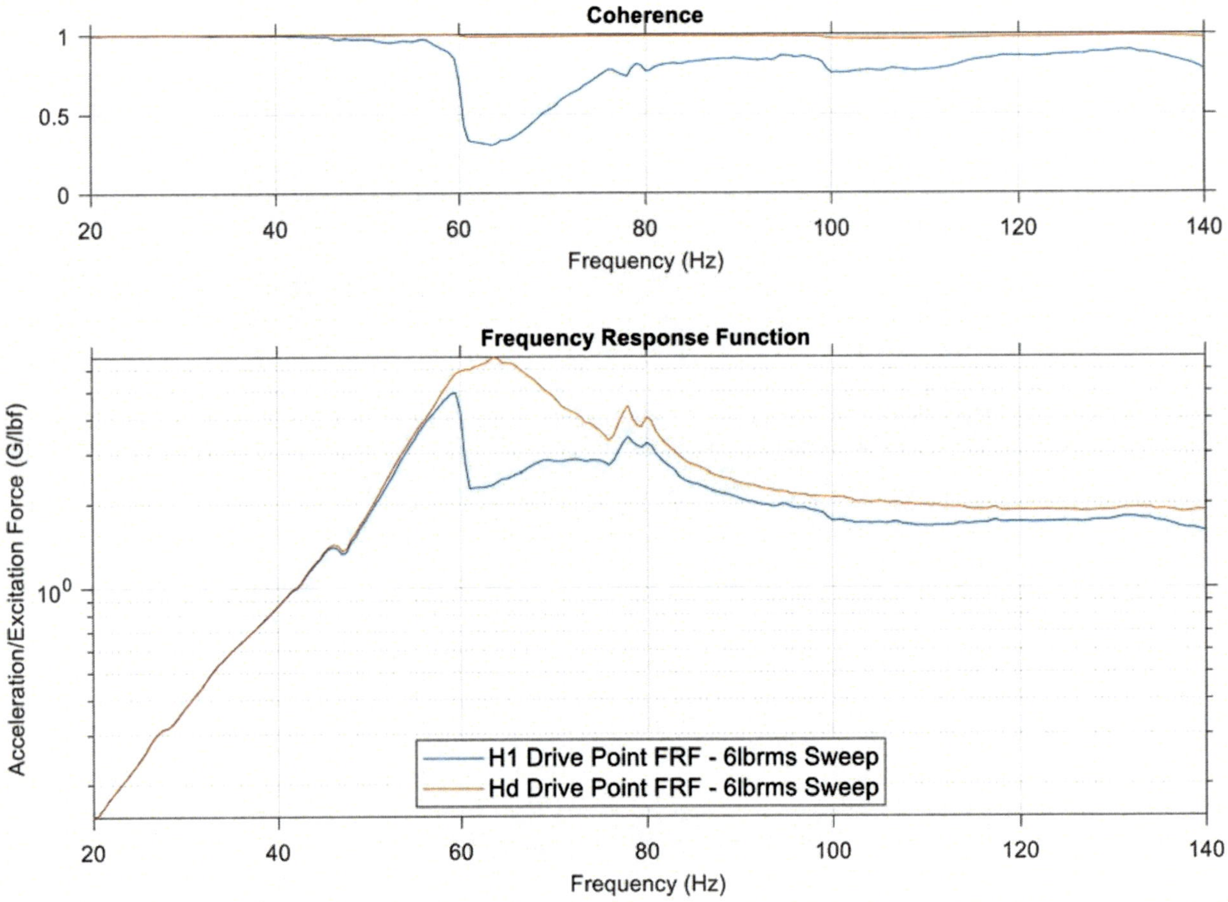

Fig. 13.3 FRF magnitude and coherence of the same data run computed using the H1 and Hd methods

unreliable FRF. To filter the out-of-band modes, the singular value decomposition method to estimate FRF [2] was used. A comparison between the H_1 and singular value decomposition FRF (H_d) is provided in Fig. 13.3.

In order to estimate the rotational dynamic stiffness of the system, the translational time-history data was converted to angular data using the geometry of the fin. The force time-history was multiplied by the chordwise distance from the CAS hinge axis to get the applied torque in units of lb-in. Likewise, the vertical acceleration time-history of the hinge axis accelerometer was subtracted from the leading-edge accelerometer location before small angle theory was applied to estimate the angular acceleration time-history of the fin. The H_d FRF of the angular acceleration relative to the applied torque was then calculated using the shaker amplifier unit voltage signal as the FRF basis. Using the input reference voltage as the basis projects the FRF onto the frequency domain encompassed by the reference voltage frequency band and filters the out-of-band modes from the acceleration response.

The locations on the fin where the excitation signal was applied and response measurements made for computation of the dynamic stiffness are shown in Fig. 13.4. Several additional accelerometers were distributed on the fin to allow for estimation of the rotational mode shape. Using the accelerometer along the root opposite, the applied load allows for a better estimate of the fin's rotation since the point is out of the load path and undergoing strain-free displacement. Likewise, subtracting the hinge vertical acceleration from the leading-edge root accelerometer accounts for any vertical rigid body deformation caused by freeplay in the CAS interface joint.

With the H_d angular acceleration FRF calculated, double integration of the FRF was conducted in the frequency domain to generate the angular displacement over applied torque FRF. For a single degree of freedom system, the low-frequency portion of the FRF corresponds to the flexibility line of the system (i.e., the inverse of the stiffness). The value of the flexibility line was computed as the average of the first 5 Hz of each respective signal. The rotation over torque FRF for each of the eight runs conducted is shown in Fig. 13.5. The calculated dynamic stiffness is listed in Table 13.2. It is important to note that the lower frequency band of the sweeps was changed between runs so as to minimize excessive shaker wear due to high amplitude, low frequency excitation force far away from the mode. The low-frequency portion of the FRF contains the most

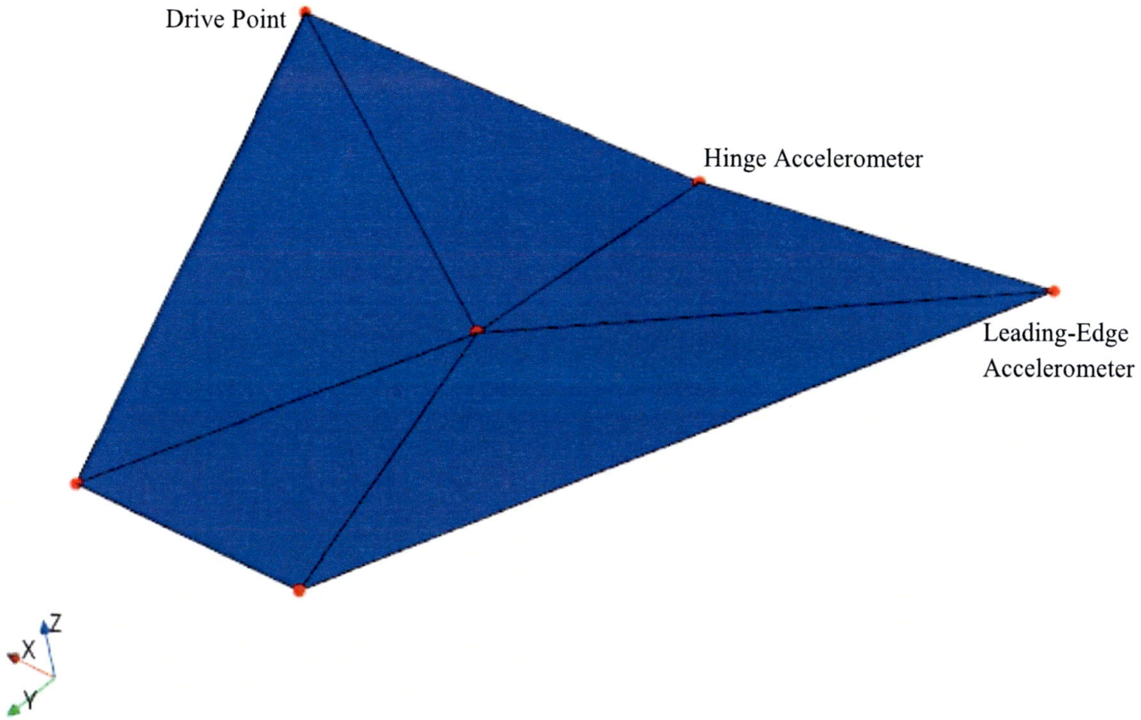

Fig. 13.4 Test display model generated for the test article showing the drive point and response locations used

accurate information related to the dynamic stiffness of the rigid body rotation, and as such, not having information at much lower frequencies than measured was a hindrance to obtaining accurate dynamic stiffness estimations, as observed by the large variation in stiffness seen in Table 13.2. This variation was particularly noticeable when the estimation range increased from 20–25 Hz to 50–55 Hz. It is thought that having FRF data down to a few Hz would result in more consistent and higher dynamic stiffness values.

The final study conducted with the dynamic test data was tracking of the frequency and damping convergence for increasing force levels. This type of study is commonly done on aircraft control surfaces and is known as CSLT. CSLT, typically, concludes when the frequency of the mode is no longer shifting relative to the applied input force level. To compute the frequency (and damping) as a function of force, the modal frequency of the rotation mode at each force level tested was estimated using the SDOFit toolkit in ATA's IMAT software. SDOFit provides the frequency and damping estimate for a single DOF system using an FRF. The mode shape was verified by animating the extracted shape at all six response locations (neglecting the drive point acceleration response) on the test display model. In this case, the drive point accelerance FRF were used to estimate the modal parameters of the rotation mode for each force level tested. Figure 13.6 shows the extracted rotation mode shape plotted on the test display model of the fin for the 2 lbrms run. The wireframe object in Fig. 13.6 is the undeformed fin, whereas the solid fin shows the displacement mode shape at the rotational resonant frequency.

The modal assurance criteria (MAC) matrix was computed for all extracted shapes to verify that the same shape was characterized for each force level. Small differences in the MAC are expected for the nonlinear system; however, the shape similarity is clearly observed in Fig. 13.7. Inspection of the mode shapes indicate that at lower force levels, the shape appears to be a pure rotation, whereas test shapes 4–8 at higher input force levels show a bending component near the load application location. This explains the difference observed in the MAC matrix, where test shapes 1–3 are very similar as are test shapes 4–8.

Once the frequency of the mode was estimated at each force level, the Fourier transform of the force time-history was used to determine the applied force level at the frequency. Table 13.3 provides the tabular data of the frequency and damping for the rotation mode at various force levels. Figure 13.8 provides a graphical representation of the same data, which indicates that the modal frequency has indeed reached convergence at the highest force level tested. This converged frequency is what is typically used as the target frequency for posttest model correlation.

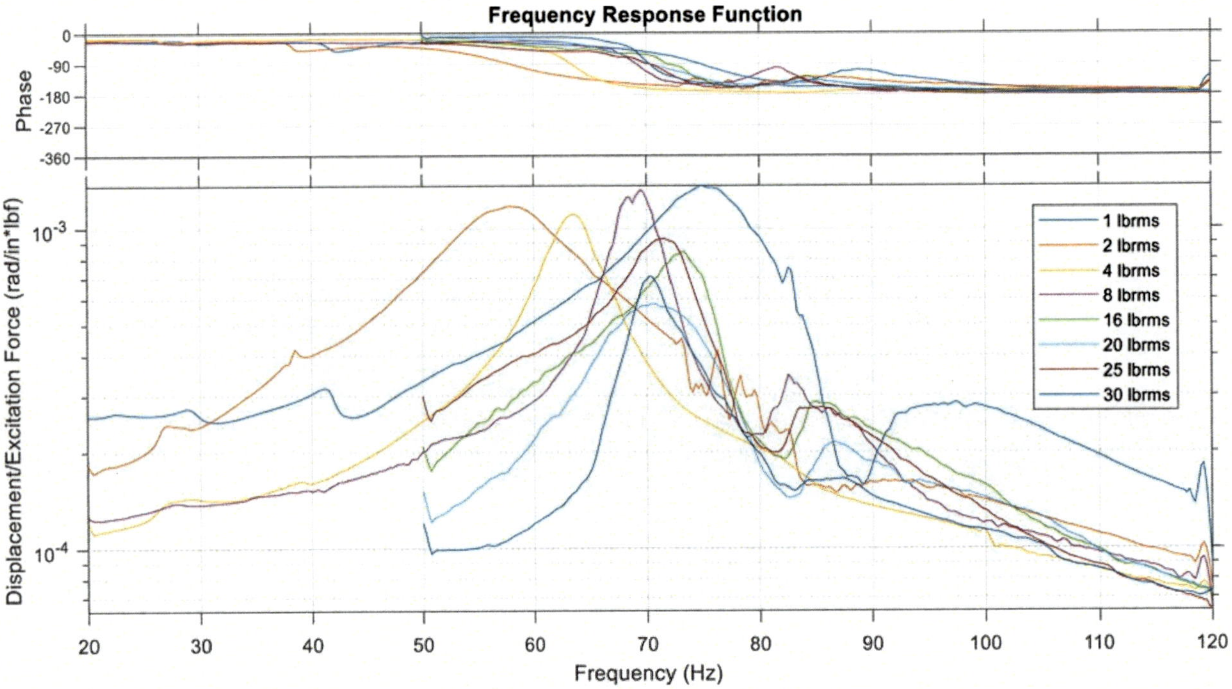

Fig. 13.5 FRF overlay for the eight nominal force levels tested

Table 13.2 Dynamic stiffness values estimated from the angular displacement FRF curves

Nominal force level (lbrms @ 120 Hz)	Rotational frequency (Hz)	Actual force at frequency (lbrms)	Rotational stiffness (in-lb/deg)	Frequency range of estimate
1	75.50	0.74	66.70	20–25 Hz
2	57.89	1.32	94.94	20–25 Hz
4	63.63	2.38	150.34	20–25 Hz
8	69.34	3.32	138.76	20–25 Hz
16	73.71	4.18	84.18	50-55 Hz
20	70.94	7.78	127.49	50–55 Hz
25	71.78	3.82	58.64	50–55 Hz
30	70.48	10.01	175.27	50–55 Hz

13.3.2 Static Testing

F&R testing was conducted on the UUT to statically characterize the stiffness of the CAS interface joint. In order to obtain F&R results, both load and displacement measurements are required to calculate the applied moment and rotation angle. A load cell was used to measure the applied load, and linear variable differential transformers (LVDTs) were used to measure the corresponding displacements of the fin. The distances from the load cell and LVDTs to the hinge line of the fin (in line with the gearbox output shaft) were used to convert load and displacement to moment and rotation angle, as was done for the dynamic tests. Transducer information and distance measurements are listed in Table 13.4. The LVDTs were mounted on an isolated assembly using tripods, and the LVDT rod ends were positioned orthogonal to the fin surface. An HBM SoMat eDAQ was used as the DAS to record the load and displacement at 200 Hz.

The load cell was positioned in series between the fin clamp and a screw-jack motor, which applied the load in a controlled manner. A ball joint was also placed in series between the load cell and the screw-jack motor to allow for purely orthogonal load inputs to the fin and to remove any potential applied moments. The motor was operated using the DAS and an ATA-developed controller specifically for F&R testing. Using the motor and controller allowed for smooth load application, consistent cycle times, and the ability to hit target hinge moments. The F&R test setup is shown in Fig. 13.9.

All displacement and load time-history data were recorded simultaneously. The test was conducted by applying five cycles of loading and unloading to the fin in both directions (positive – load cell tension and negative – load cell compression). After recording the measured data, the bending displacement (measured at LVDT 4 on the rotation hinge line) was subtracted

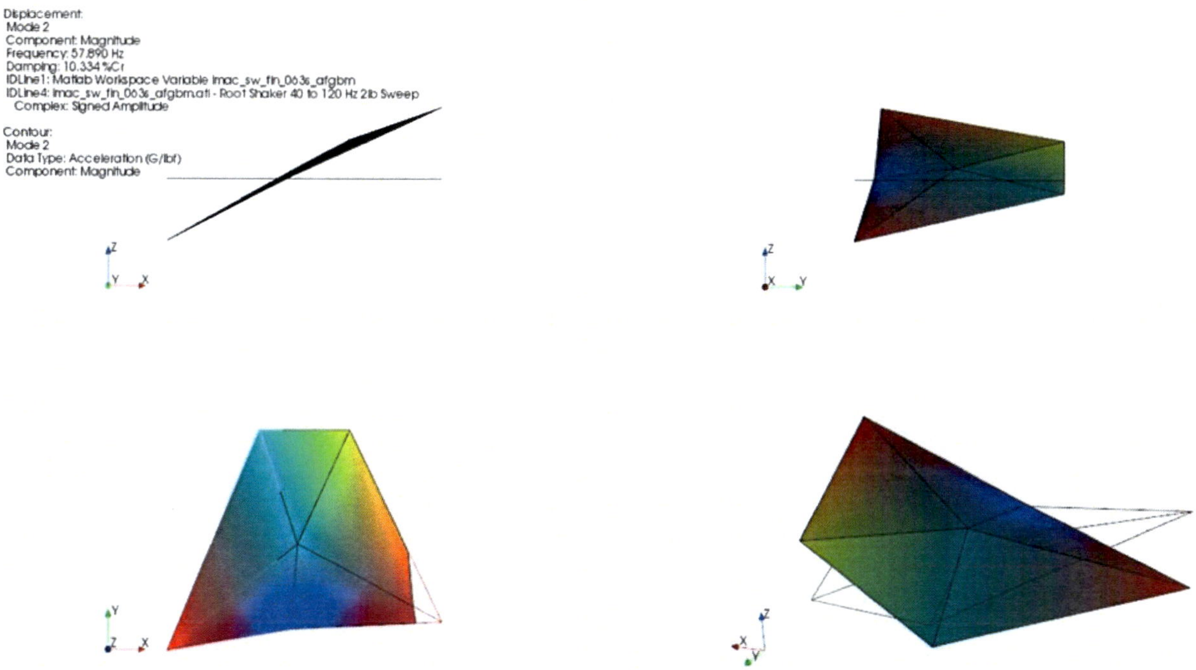

Fig. 13.6 Mode shape plot for the extracted rotation mode plotted on the test display model

Test Self MAC Table

			Test Shapes							
			1	2	3	4	5	6	7	8
MACtt	**Test Shapes**		75.50	57.89	63.63	69.34	73.71	70.93	71.78	70.48
100	1	75.50	100	97	98	83	90	96	89	93
95	2	57.89	97	100	100	73	78	87	78	84
90	3	63.63	98	100	100	78	82	89	82	88
85	4	69.34	83	73	78	100	97	91	98	97
80	5	73.71	90	78	82	97	100	98	100	99
75	6	70.93	96	87	89	91	98	100	97	98
70	7	71.78	89	78	82	98	100	97	100	99
65	8	70.48	93	84	88	97	99	98	99	100

Fig. 13.7 MAC of the rotation modes extracted for the various force level tested

Table 13.3 Frequency and damping of rotation modes as a function of applied force level

Nominal Set Force at 120 Hz (lbf-rms)	Force at Frequency (lbf-rms)	Frequency (Hz)	Damping (%Cr)	Force %Diff	Frequency %Diff
1	0.74	75.50	6.62		
2	1.32	57.58	10.33	79.6%	−23.7%
4	2.38	63.58	3.84	79.6%	10.4%
8	3.32	69.31	2.89	39.9%	9.0%
25	3.82	71.72	4.08	14.9%	3.5%
16	4.18	73.67	3.50	9.4%	2.7%
20	7.78	70.64	9.07	86.2%	−4.1%
30	10.01	70.44	3.24	28.7%	−0.3%

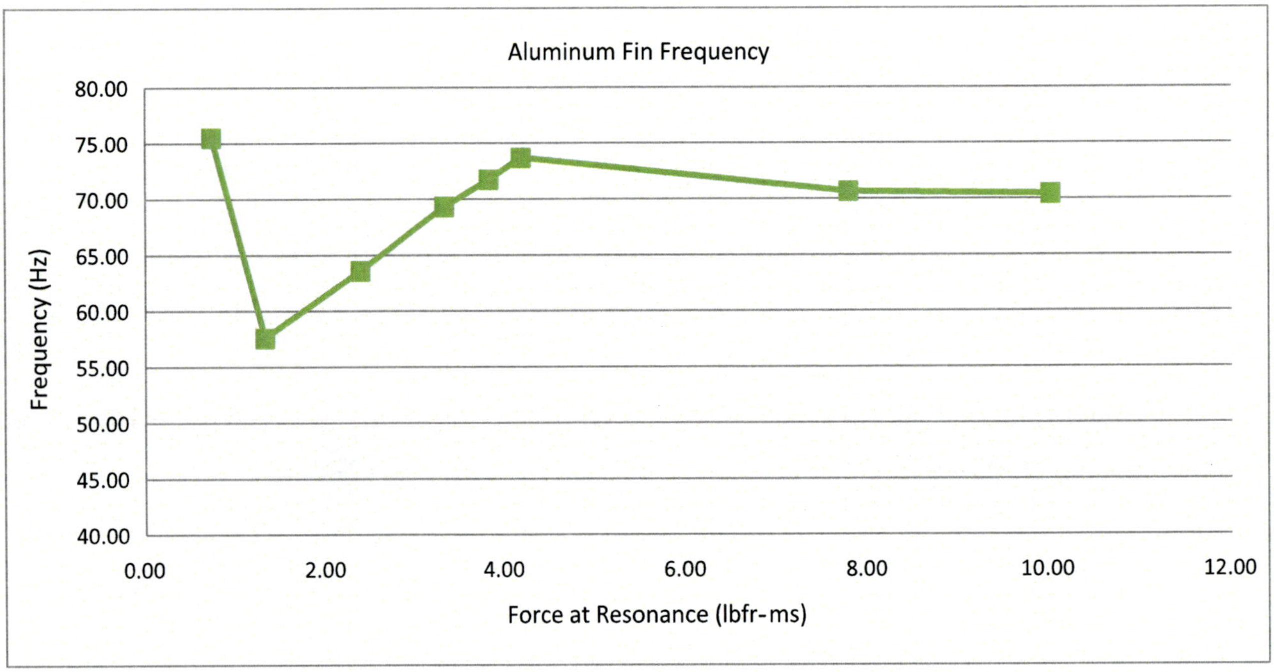

Fig. 13.8 Frequency of the rotation mode plotted as a function of applied force at the modal frequency

Table 13.4 List of the transducers used for the static testing

Transducer	Distance from Hinge Line (inch)	Model	Nominal sensitivity
Load cell	4.75	Interface 1500ASL-300	10 mV/lb
LVDT 1	4.25	Omega LD620-25	5000 mV/in
LVDT 2	4.50	Omega LD620-25	5000 mV/in
LVDT 3	2.00	Omega LD620-25	5000 mV/in
LVDT 4	0.00	Omega LD620-25	5000 mV/in

from the other LVDT locations to remove any bending of the test setup from the rotation displacement measurements. The displacement and load values were, respectively, converted into angular deflection (using the relative displacement between LVDT 1 and LVDT 3) and hinge moment. The moment-versus-rotation time-history plot for all cycles is shown in Fig. 13.10, and the hysteresis overlay for the three cycles used in the results is shown in Fig. 13.11. These plots show that the measurements were repeatable across cycles. A linear regression was performed on the upper and lower loading regions for each cycle in order to compute F&R values in accordance to JSSG-2006 [3] standards, as shown in Fig. 13.12. Figure 13.13 shows the linear fits and F&R results for cycle 3. The results for cycles 3, 4, and 5 are summarized in Table 13.5.

Table 13.5 shows consistent results across the three cycles used in the computation. The shaded region near the tips of the positive and negative loading cycles indicate the region from which the rigidity (stiffness) was estimated. These regions were chosen as they are responding linearly, as indicated by the dashed lines in Fig. 13.13. The consistency between cycles and also between positive and negative loading provides high confidence in the static stiffness estimates.

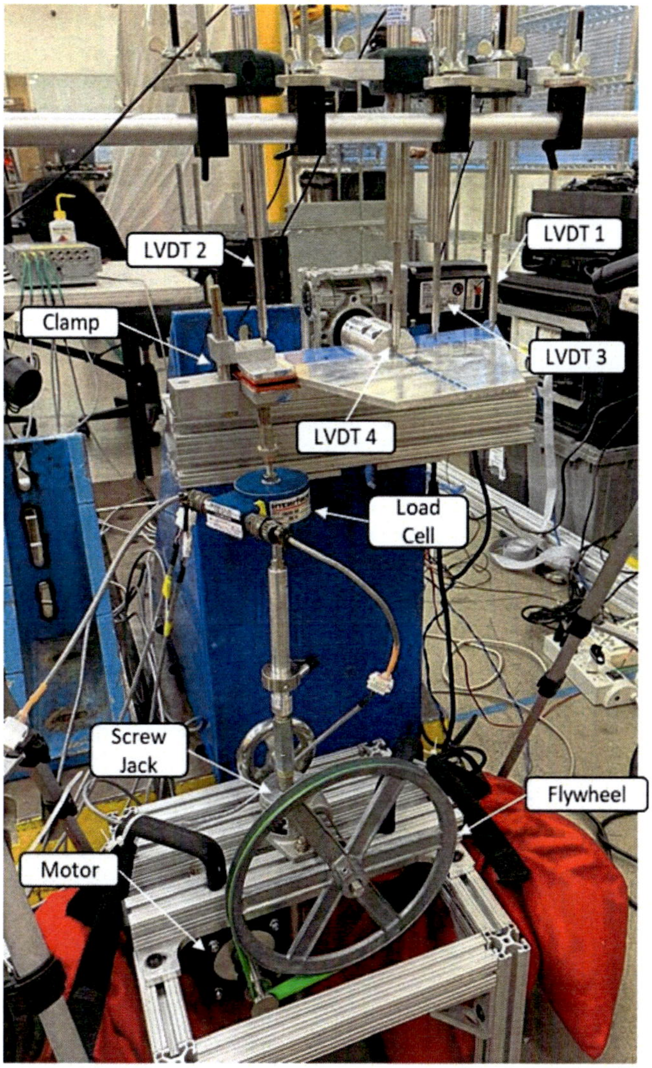

Fig. 13.9 F&R test setup

13.4 Conclusion

Static and dynamic testing was conducted on a missile fin CAS testbed to characterize the stiffness of the CAS interface joint. During dynamic testing, the rotational frequency of the fin was found to converge with respect to applied force at the modal frequency. This converged frequency was 70.5 Hz, with an estimated hinge stiffness of 175.27 in-lb/deg. at 10 lbrms. The stiffness estimates from the dynamic testing captured the general trend of increasing stiffness with increasing applied force level; however, a lack of low frequency information jeopardizes the accuracy of the stiffness estimates. It is recommended that if dynamic stiffness estimates from the FRF are desired, data are taken as low in frequency as practicable.

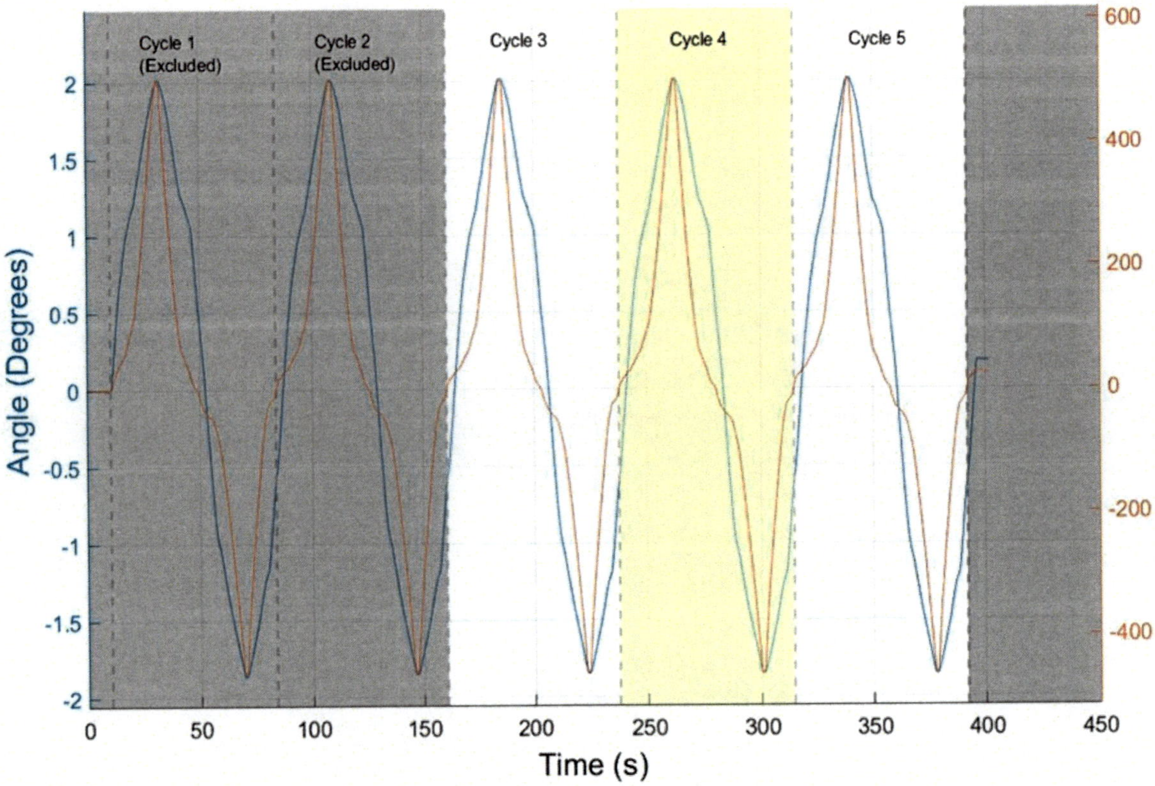

Fig. 13.10 Hinge moment and angle time-history

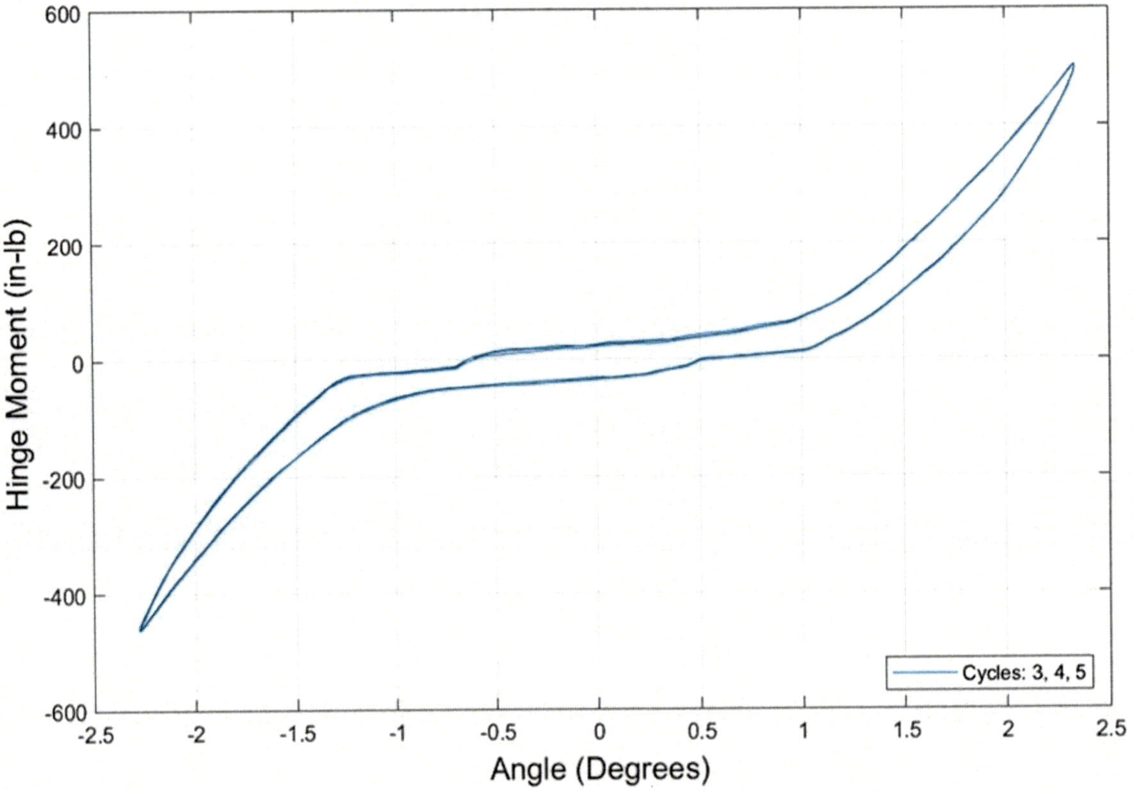

Fig. 13.11 Hysteresis overlay of cycles 3, 4, and 5

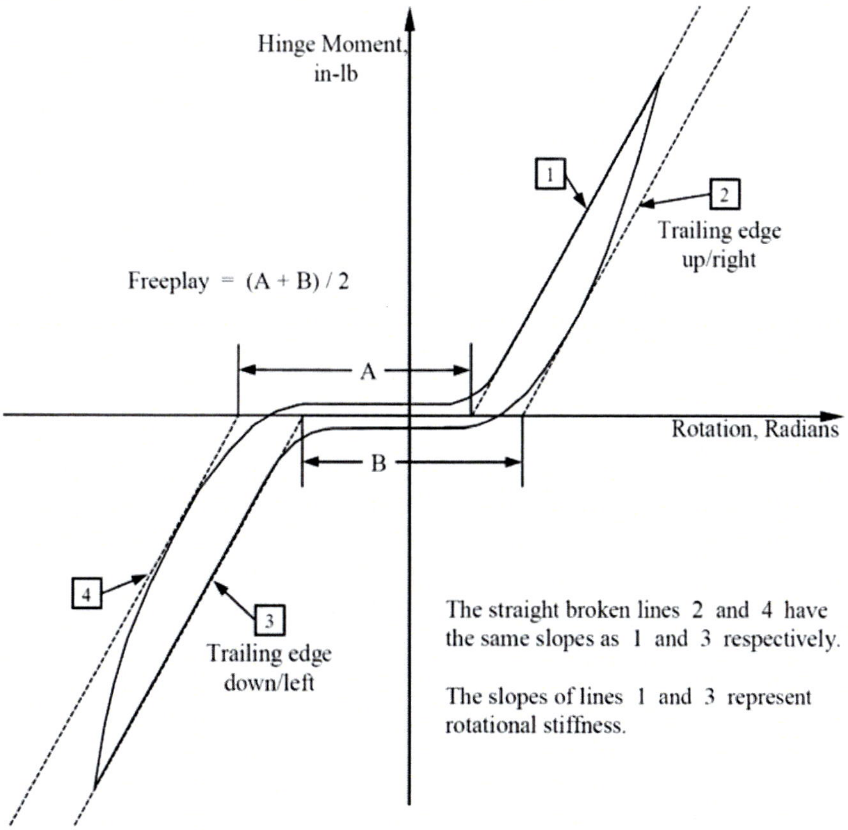

Fig. 13.12 Calculations per JSSG-2006 for freeplay and rigidity

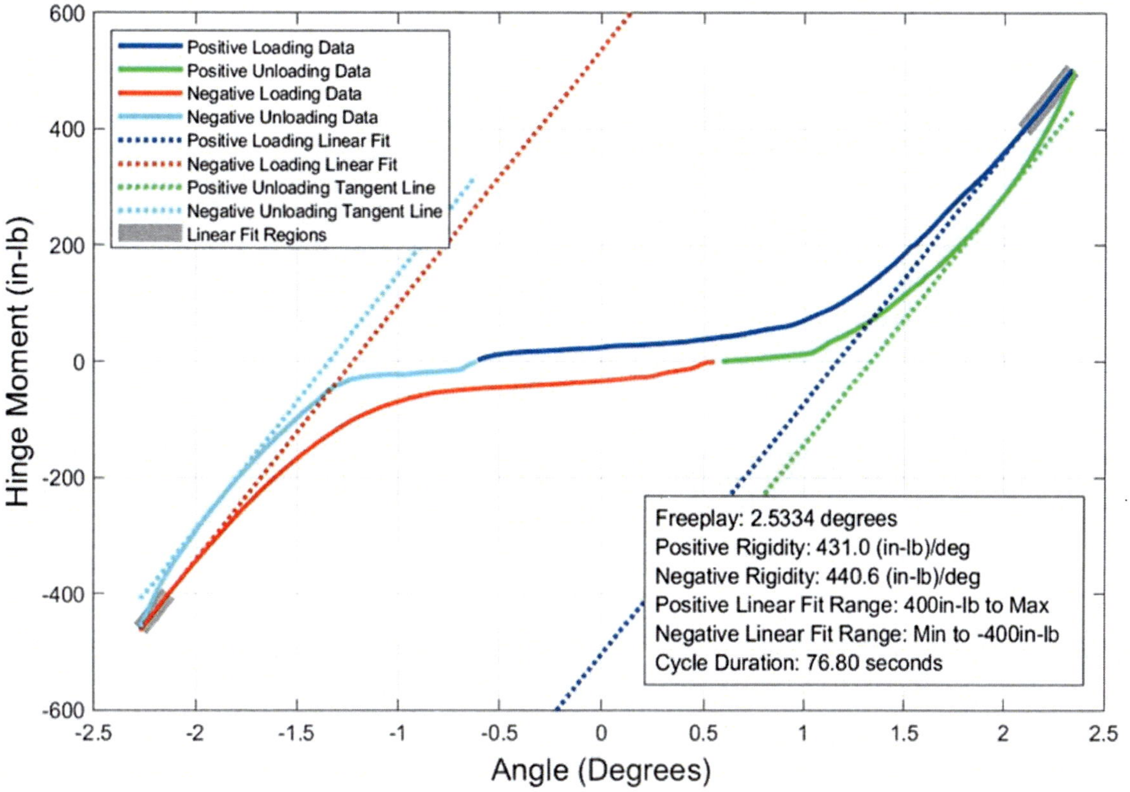

Fig. 13.13 Hinge moment versus angle for cycle 3

Table 13.5 F&R results

Cycle	Freeplay (degrees)	Positive rigidity (in-lb/deg)	Negative rigidity (in-lb/deg)	Cycle duration (sec)	Lower bound (in-lbf)	Upper bound (in-lbf)
3	2.53	431.00	440.62	76.80	−400	400
4	2.56	440.57	442.46	76.40		
5	2.59	436.99	465.26	77.00		
Average	*2.56*	*436.19*	*449.45*	*76.73*		

The static tests revealed the hinge stiffness to be highly nonlinear at low force levels, with the stiffness not reaching a linear profile until approximately 350 in-lb of applied torque. The stiffness of the hinge at these higher force levels was estimated to be 430–440 in-lb/deg., 2.5 times higher than the hinge stiffness recovered from the dynamic tests. These results suggest that the standard practice of performing CSLT and using the converged frequency value for model correlation and flutter analysis may discount the linear CAS interface joint stiffness.

References

1. Waszak, M.R., Buttrill, C.S., Schmidt, D.K.: Modeling and Model Simplification of Aeroelastic Vehicles: An Overview. NASA Technical Memorandum (1992)
2. Napolitano, K.: Using Singular Value Decomposition to Estimate Frequency Response Functions. In: Proceedings, 34th International Modal Analysis Conference (2016)
3. JSSG-2006: DOD Joint Service Specification Guide: Aircraft Structures (1998)

Chapter 14
Dynamic Characterization of Aircraft Shock Cords Used for Free-Free Boundary Conditions for Ground Vibration Testing

Joseph M. Jaeckels, Arthur J. Nguyen, and Douglas J. Osterholt

Abstract In ground vibration tests (GVT), it is often critical to separate rigid body modes from flexible body modes to simulate a free-free boundary condition. This is often done by chaining several aircraft shock cords in series and parallel to achieve a stiffness that results in a sufficiently isolated rigid body bounce mode for the test article. Thus, the stiffness values of these aircraft shock cords are critical in estimating the rigid body bounce frequency. However, the stiffness of these aircraft shock cords are often uncharacterized by the manufacturer or are presented as an elongation percentage at a given static load. One approach to characterizing the stiffness of these shock cords is to measure deflection as a function of an applied static load. Using this approach has typically resulted in significant underestimation of the measured rigid body bounce frequency, potentially leading to less than desirable separation between rigid body modes and flexible body modes. In response to this discrepancy, ATA performed an in-depth study to better characterize the dynamic stiffness of these aircraft shock cords to better estimate rigid body isolation frequencies. This chapter presents the methodology used to characterize the dynamic stiffness of these shock cords and the results of this study, which has significantly reduced bounce mode frequency estimation error.

Keywords Modal testing · Ground vibration testing · Rigid body modes · Soft-suspension system · Aircraft shock cord

Nomenclature

ATA ATA Engineering, Inc.
GVT Ground vibration test

14.1 Introduction

Ground vibration tests (GVTs) are often conducted with a simulated free-free boundary condition [1–3] on the test article in order to isolate rigid body modes from flexible body modes. This boundary condition can be approximated with a variety of methods, such as deflating landing gear tires, resting the test article on inflatable airbags or canisters, or suspending the test article with an elastic suspension cord. The boundary conditions play an important role when performing model updates after the test [4]. ATA Engineering (ATA) typically uses aircraft shock cords placed in series and in parallel to create a suspension cord with a predicted stiffness that would isolate the rigid bounce mode from the first flexible mode. However, the dynamic stiffness of these shock cords is often uncharacterized by the manufacturers. This unknown dynamic stiffness can result in underestimation of measured rigid body bounce frequency, resulting in less than adequate separation between rigid and flexible test article modes.

Manufacturers often characterize aircraft shock cords with a maximum load rating or as an elongation percentage at a given force. Historically, ATA has measured the static stiffness of these shock cords by measuring elongation along various load points, resulting in a static-stiffness curve for a given shock cord type. ATA has then used these static-stiffness curves to predict test article suspension bounce mode frequencies. However, ATA has noticed significant discrepancies between

J. M. Jaeckels · A. J. Nguyen · D. J. Osterholt (✉)
ATA Engineering, Inc., San Diego, CA, USA
e-mail: doug.osterholt@ata-e.com

C. Walber et al. (eds.), *Sensors & Instrumentation and Aircraft/Aerospace Testing Techniques, Volume 8*, Conference Proceedings of the Society for Experimental Mechanics Series, https://doi.org/10.1007/978-3-031-34938-6_14

Table 14.1 Measured (from prior GVTs) and predicted (from static-stiffness estimates) bounce frequencies

Test article weight (lbf)	Shock cord suspension setup	Measured bounce frequency (Hz)	Estimated bounce frequency (Hz)	Percent difference
3798	8 in parallel, 3 in series	1.03	0.59	−42.3%
7431	11 in parallel, 2 in series	1.35	1.08	−20.0%
12,846	19 in parallel, 2 in series	1.31	1.08	−17.6%
10,515	17 in parallel, 2 in series	1.50	1.12	−25.3%
10,061	16 in parallel, 2 in series	1.36	1.12	−17.7%
9222	15 in parallel, 2 in series	1.35	1.13	−16.3%
2711.5	4 in parallel, 3 in series	1.22	0.62	−49.2%
3112	4 in parallel, 3 in series	1.34	0.79	−41.0%
2653	4 in parallel, 2 in series	1.10	0.63	−42.7%
3038	4 in parallel, 3 in series	1.22	0.66	−45.9%
2777	4 in parallel, 3 in series	1.00	0.61	−39.0%

the predicted and measured bounce modes in various past GVTs. These discrepancies are summarized in Table 14.1, which compares measured bounce mode frequencies (from various test article GVTs conducted by ATA) to estimated bounce frequencies (computed from shock cord static-stiffness estimates). In order to better understand these discrepancies and have more accurate future GVT suspension predictions, ATA investigated the dynamic characterization of these shock cords. The scope of this paper is limited to D1INT23 (1″ thick cord, 23″ internal diameter) aircraft shock cords.

14.2 Experimental Setup

Two test setups were utilized to fully characterize the static- and dynamic-stiffness estimates of the aircraft shock cord. The first test setup was used to estimate static-stiffness values of the shock cord by measuring its elongation as a function of applied force. A single shock cord was secured to a reaction floor, applied force was measured with a load cell, displacement was measured with a string potentiometer, and force was applied by an overhead crane. Three shock cords were also tested simultaneously, in order to measure changes in stiffness due to parallel loading. The static test setup with multiple shock cords in parallel is shown in Fig. 14.1.

The second test setup estimated bounce frequency (dynamic stiffness) at various preload weights. For this test, a load cell was placed in-line of the suspended steel mass to measure applied force on the aircraft shock cord, while accelerations were measured with accelerometers mounted on the steel mass and shackles above and below the shock cord. Excitation was provided to the system by manually exciting the bounce mode of the hanging mass and measuring the response force and accelerations. The test setup is shown in Fig. 14.2.

14.3 Results

Static-stiffness values were determined by computing the slope of the measured force-versus-displacement plots for the single-cord static test setup. The results of the static study are summarized in Table 14.2. The static-stiffness curve for the D1INT23 shock cord is presented in Fig. 14.3.

Dynamic-stiffness estimates were computed using two different estimation methods. The first estimate was computed with the measured frequency from accelerometer time history recordings, using Eq. (14.1). The second estimate was computed by integrating the recorded accelerations of the bouncing mass into displacements and computing the slope of a measured force-vs.-displacement curve for a given load point. Results of the two approaches are summarized in Table 14.3, with plots of the dynamic stiffness as a function of applied load shown in Fig. 14.4.

$$k = \text{Dynamic stiffness} = M \times (2\pi f_{\text{measured}})^2 \tag{14.1}$$

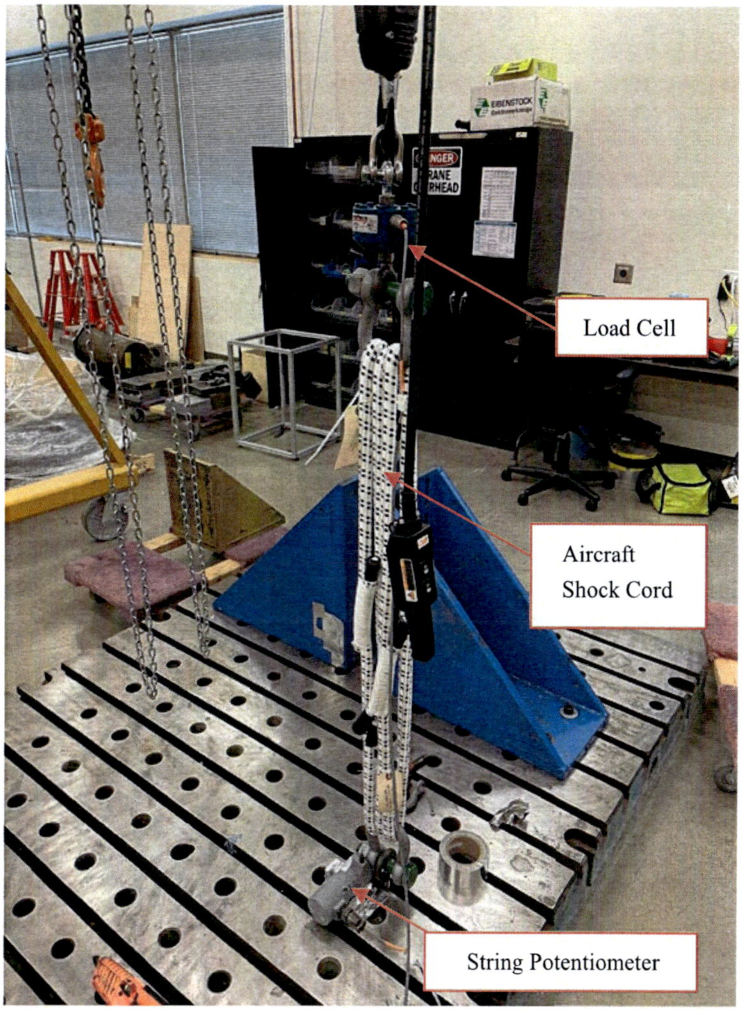

Fig. 14.1 Shock cord static-stiffness test setup

$$M = \text{Applied mass}$$

$$f_{\text{measured}} = \text{Measured bounce frequency}$$

A limited-case dynamic-stiffness study was conducted with three shock cords in parallel to assess any influence parallel loading may have on measured bounce frequency. The results comparing the parallel shock cords to a single shock cord are summarized in Table 14.4, where both dynamic-stiffness values were computed using the measured bounce frequency. Although noticeable dynamic-stiffness change was noted with parallel shock cords, this change translated to an insignificant change in bounce frequency, as summarized in Table 14.5, which compares the parallel shock cord bounce frequency to the bounce frequency of a single shock cord at equivalent applied load.

In order to assess the accuracy of bounce mode frequency estimation using static and dynamic-stiffness values, a theoretical bounce frequency was computed using both the static-stiffness values and the dynamic-stiffness values computed from measured loads. These predictions were compared to the measured bounce frequency and are summarized in Table 14.6, with a plot overlaying all three frequency values in Fig. 14.5. Note that using the dynamic-stiffness estimates to compute bounce frequency resulted in significantly more accurate bounce frequency estimates compared to values computed from the static stiffness.

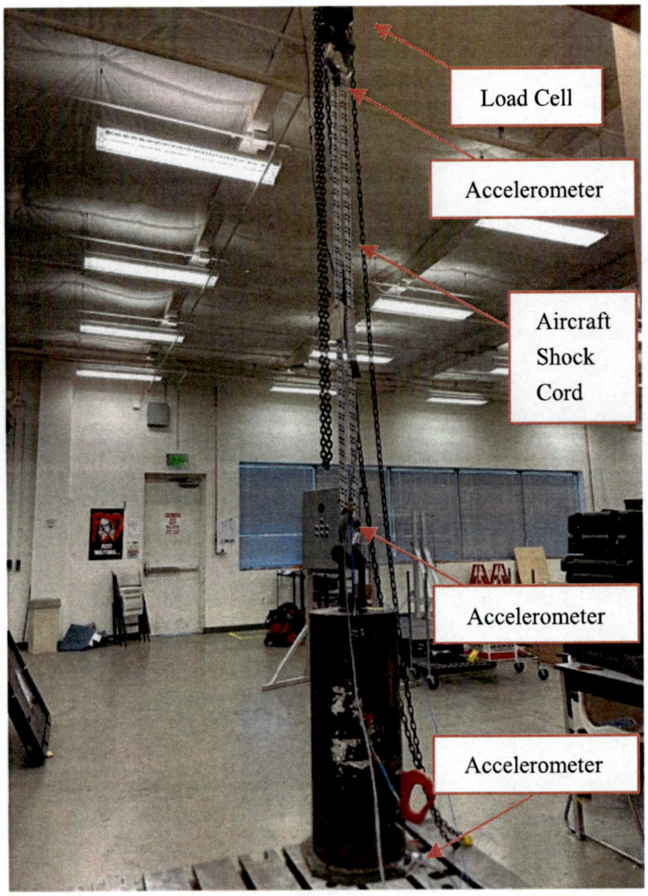

Fig. 14.2 Bounce mode characterization test setup

Table 14.2 D1INT23 shock cord static-stiffness test results

Load (lbf)	Displacement (in)	% elongation	Total length (in)	Single bungee ± 25 lbf. linear fit avg. static stiffness (lbf/in)
375	3.87	11%	37.87	23.39
400	4.74	14%	38.74	23.62
425	5.76	17%	39.76	21.09
450	6.90	20%	40.90	20.34
475	8.11	24%	42.11	21.19
500	9.33	27%	43.33	21.82
525	10.50	31%	44.50	22.57
550	11.61	34%	45.61	23.83
575	12.69	37%	46.69	25.90
600	13.74	40%	47.74	26.01
625	14.72	43%	48.72	28.18
650	15.64	46%	49.64	29.84
675	16.51	49%	50.51	32.25
700	17.32	51%	51.32	33.41
725	18.09	53%	52.09	33.26
750	18.85	55%	52.85	34.65
775	19.73	58%	53.73	33.70

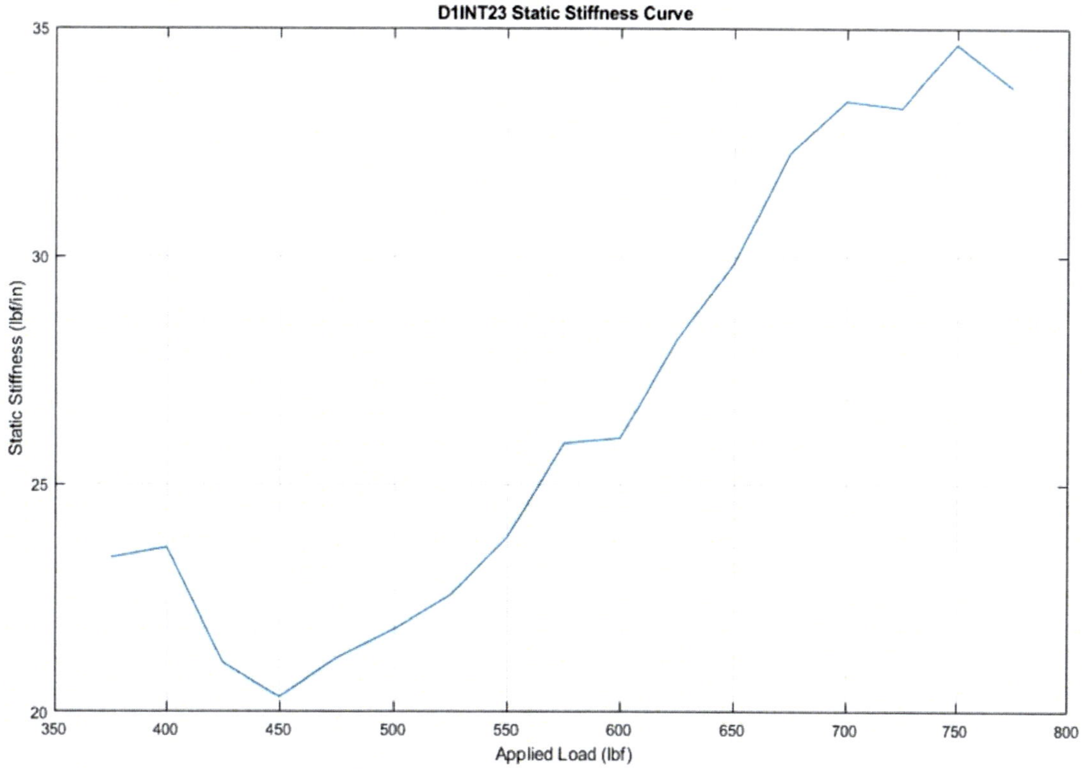

Fig. 14.3 D1INT23 static-stiffness curve

Table 14.3 D1INT23 shock cord dynamic-stiffness test results

Load (lbf)	Measured frequency (Hz)	Dynamic stiffness using measured frequency (lbf/in)	Dynamic stiffness using measured force (lbf/in)
375	1.01	39.1	40.3
400	0.90	33.4	35.2
425	0.83	30.1	32.3
450	0.81	30.0	32.5
475	0.78	29.7	33.0
500	0.76	29.3	33.6
525	0.76	31.3	36.3
550	0.76	32.3	37.8
575	0.77	35.3	40.5
600	0.81	40.4	44.2
625	0.85	46.2	50.0
650	0.86	49.3	52.9
675	0.88	53.3	57.2
700	0.92	60.4	62.5
725	0.94	65.1	67.2
750	0.98	73.2	75.7
775	1.01	81.3	80.4

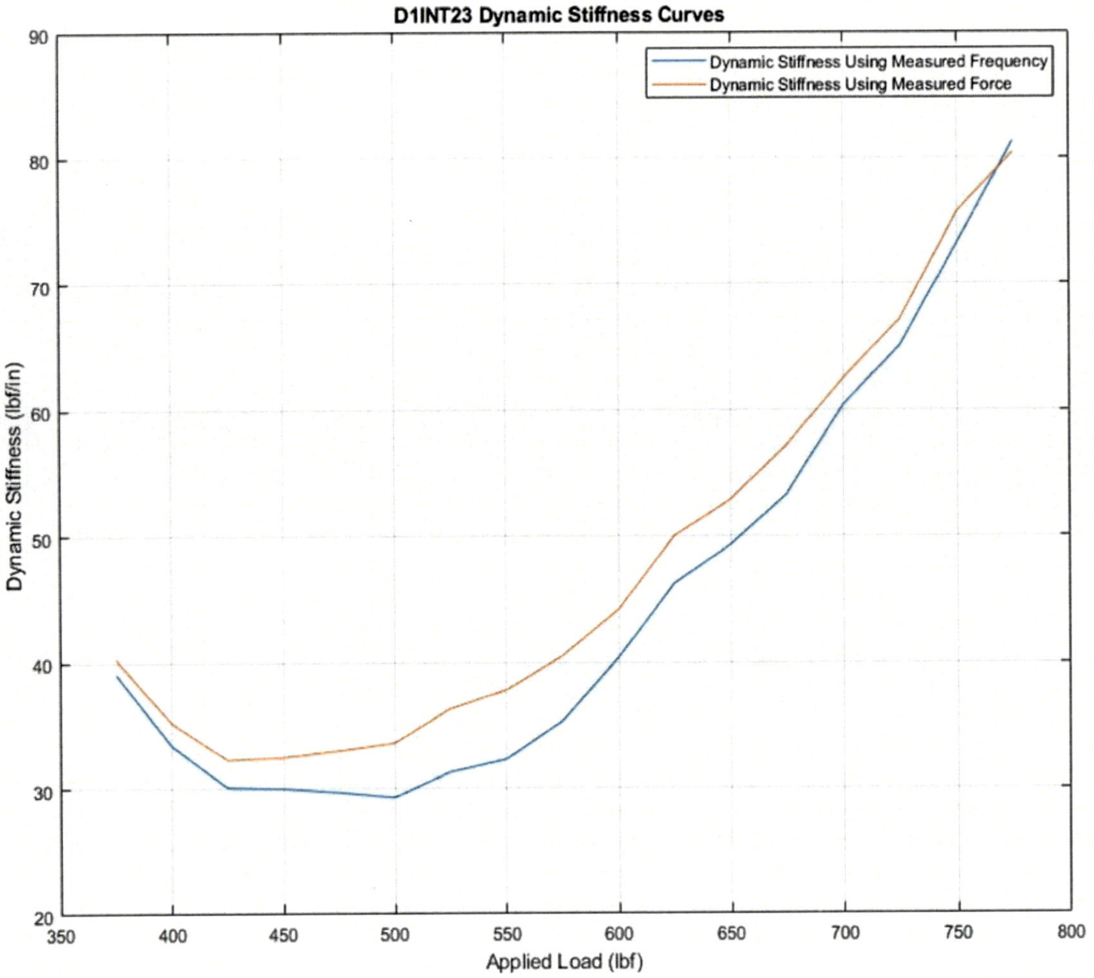

Fig. 14.4 D1INT23 dynamic-stiffness curve

Table 14.4 Parallel shock cord dynamic-stiffness test results

Total load (lbf)	Equivalent load per shock cord (lbf)	Measured frequency (Hz)	Dynamic stiffness using measured frequency (lbf/in)	Single Bungee dynamic stiffness at equivalent applied load (lbf/in)	Percent difference (parallel shock cord vs. single shock cord)
1500	500	0.80	32.6	29.3	+11.4%
1575	525	0.79	33.4	31.3	+6.6%
1650	550	0.79	35.0	32.3	+8.2%
1725	575	0.79	36.5	35.3	+3.3%
1800	600	0.79	39.3	40.4	−2.7%

Table 14.5 Parallel shock cord bounce frequency comparison

Total load (lbf)	Equivalent load per shock cord (lbf)	Measured frequency (Hz)	Measured frequency of single shock cord at equivalent load (Hz)	Frequency percent difference (parallel shock cord vs. single shock cord)
1500	500	0.80	0.76	+5.5%
1575	525	0.79	0.76	+3.2%
1650	550	0.79	0.76	+4.0%
1725	575	0.79	0.77	+1.6%
1800	600	0.79	0.81	−1.3%

Table 14.6 Frequency estimation using static- and dynamic-stiffness comparison

Load (lbf)	Measured frequency (Hz)	Computed frequency using static stiffness (Hz)	Percent difference to measured frequency	Computed frequency using dynamic stiffness from measured force (Hz)	Percent difference to measured frequency
375	1.01	0.81	−19.7%	1.00	−0.8%
400	0.90	0.74	−17.9%	0.91	+0.1%
425	0.83	0.68	−18.2%	0.84	+1.1%
450	0.81	0.65	−19.6%	0.82	+1.7%
475	0.78	0.64	−17.5%	0.80	+2.9%
500	0.76	0.64	−15.7%	0.79	+4.6%
525	0.76	0.63	−17.1%	0.80	+5.1%
550	0.76	0.64	−16.2%	0.80	+5.6%
575	0.77	0.65	−16.4%	0.81	+4.6%
600	0.81	0.64	−21.7%	0.83	+2.1%
625	0.85	0.65	−23.7%	0.86	+1.6%
650	0.86	0.65	−24.0%	0.87	+1.1%
675	0.88	0.67	−24.0%	0.89	+1.2%
700	0.92	0.67	−27.4%	0.91	−0.7%
725	0.94	0.65	−30.2%	0.93	−0.8%
750	0.98	0.66	−32.8%	0.97	−0.7%
775	1.01	0.64	−37.1%	0.98	−2.9%

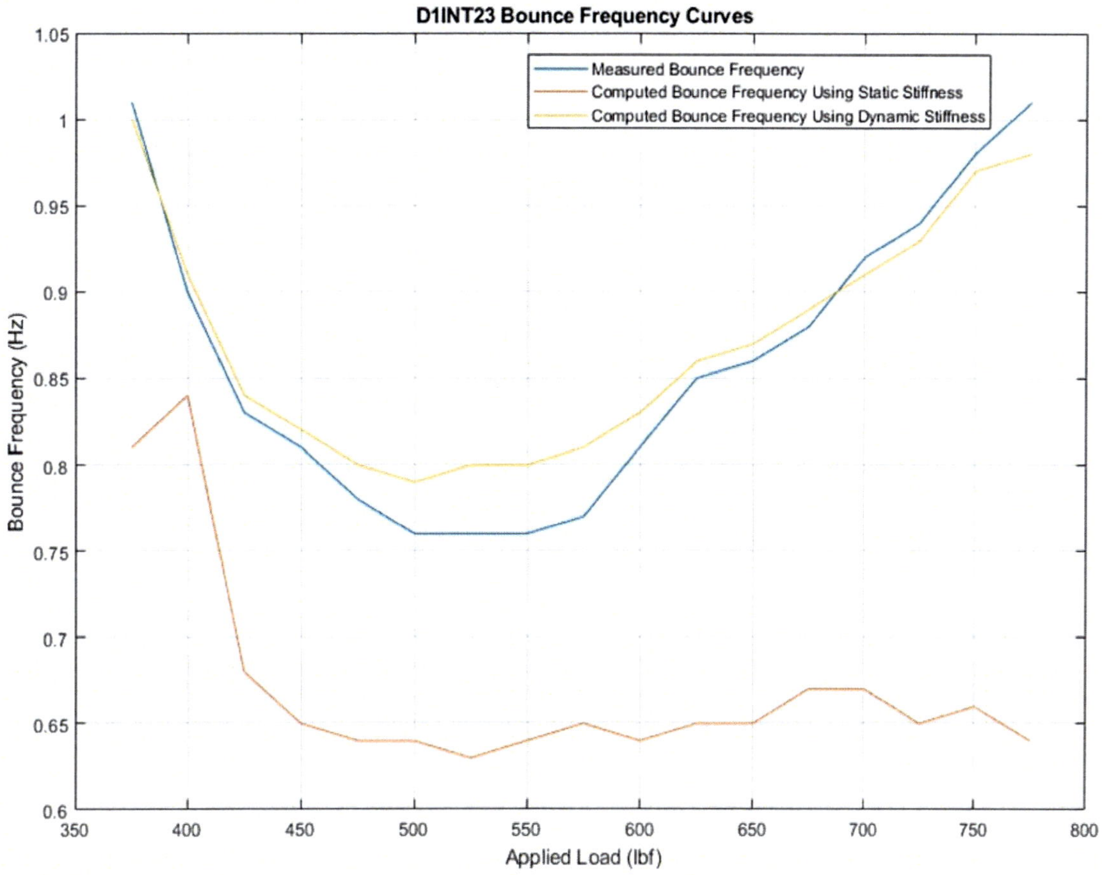

Fig. 14.5 D1INT23 bounce frequency comparison curves

Table 14.7 Comparison of measured (from prior GVTs) and predicted (from dynamic-stiffness estimates) bounce frequencies

Test article weight (lbf)	Shock cord suspension setup	Measured bounce frequency (Hz)	Estimated bounce frequency from static stiffness (Hz)	Percent difference (Static stiffness to measured frequency)	Estimated bounce frequency from static stiffness (Hz)	Percent difference (Dynamic stiffness to measured frequency)
3798	8 in parallel, 3 in series	1.03	0.59	−42.3%	0.91	−11.7%
7431	11 in parallel, 2 in series	1.35	1.08	−20.0%	1.40	+3.4%
12,846	19 in parallel, 2 in series	1.31	1.08	−17.6%	1.40	+6.6%
10,515	17 in parallel, 2 in series	1.50	1.12	−25.3%	1.32	−12.2%
10,061	16 in parallel, 2 in series	1.36	1.12	−17.7%	1.33	−2.2%
9222	15 in parallel, 2 in series	1.35	1.13	−16.3%	1.31	−3.1%
2711.5	4 in parallel, 3 in series	1.22	0.62	−49.2%	1.14	−6.4%
3112	4 in parallel, 3 in series	1.34	0.79	−41.0%	1.50	+12.3%
2653	4 in parallel, 2 in series	1.10	0.63	−42.7%	1.12	+1.8%
3038	4 in parallel, 3 in series	1.22	0.66	−45.9%	1.22	−0.1%
2777	4 in parallel, 3 in series	1.00	0.61	−39.0%	1.16	+15.8%

This test methodology was repeated with other shock cord models similar to those used in prior ATA-conducted GVTs, and the new dynamic-stiffness results were then used to recalculate bounce frequency estimates of past GVT test articles. The results of this comparison are presented in Table 14.7, which compares the measured bounce frequency of past GVTs and their original estimates derived from static deflection data with the new recomputed bounce frequency using dynamic-stiffness data. Note that the dynamic estimates yielded much more accurate estimates when compared to the older static estimates, though there may be additional error associated with the fact that the shock cords used in the recent dynamic characterization effort were not the exact batch or make of the shock cords used in the original GVTs but are equivalent shock cords of similar specification. An example of the spreadsheet that utilizes this dynamic-stiffness data to predict future GVT bounce frequencies is shown in Fig. 14.6.

14.4 Conclusion

An in-depth characterization of aircraft shock cords revealed that using stiffness values derived from static displacements yielded bounce estimates that were significantly lower than measured bounce mode frequencies. Bounce mode frequencies computed with dynamic data yielded bounce mode estimates that were much closer to measured bounce frequencies. Accurate bounce mode estimation is required in order to ensure adequate separation between rigid body modes and flexible body modes. The dynamic-stiffness curves can be used to optimize the number of bungees required for the test article weights. For example, either the bounce frequency could be minimized or the required number of shock cords could be minimized, depending on which is more important. Future studies are planned to investigate how different excitation methods (e.g., sine vs. random) and variations in the amplitude of excitation affect the bounce frequencies.

inputs in orange	Config 1	Config 2	Config 3	Config 4	Config 5
Shock Cord Type:	1380HD				
Recommended Cords in Parallel:	10 - 13	11 - 14	12 - 15	13 - 17	17 - 22
Number of Loops in Parallel:	12	12	12	17	17
Number of Loops in Series:	3				
Test Article Weight (lbf):	4700	5000	5300	6000	7700
Weight per bungee (lbf):	392	417	442	353	453
Estimated Bounce Mode (Hz):	0.88	0.89	0.90	0.87	0.90
Estimated Bounce Mode Range (Hz):	0.80 - 0.97	0.80 - 0.98	0.81 - 0.99	0.78 - 0.95	0.81 - 0.99

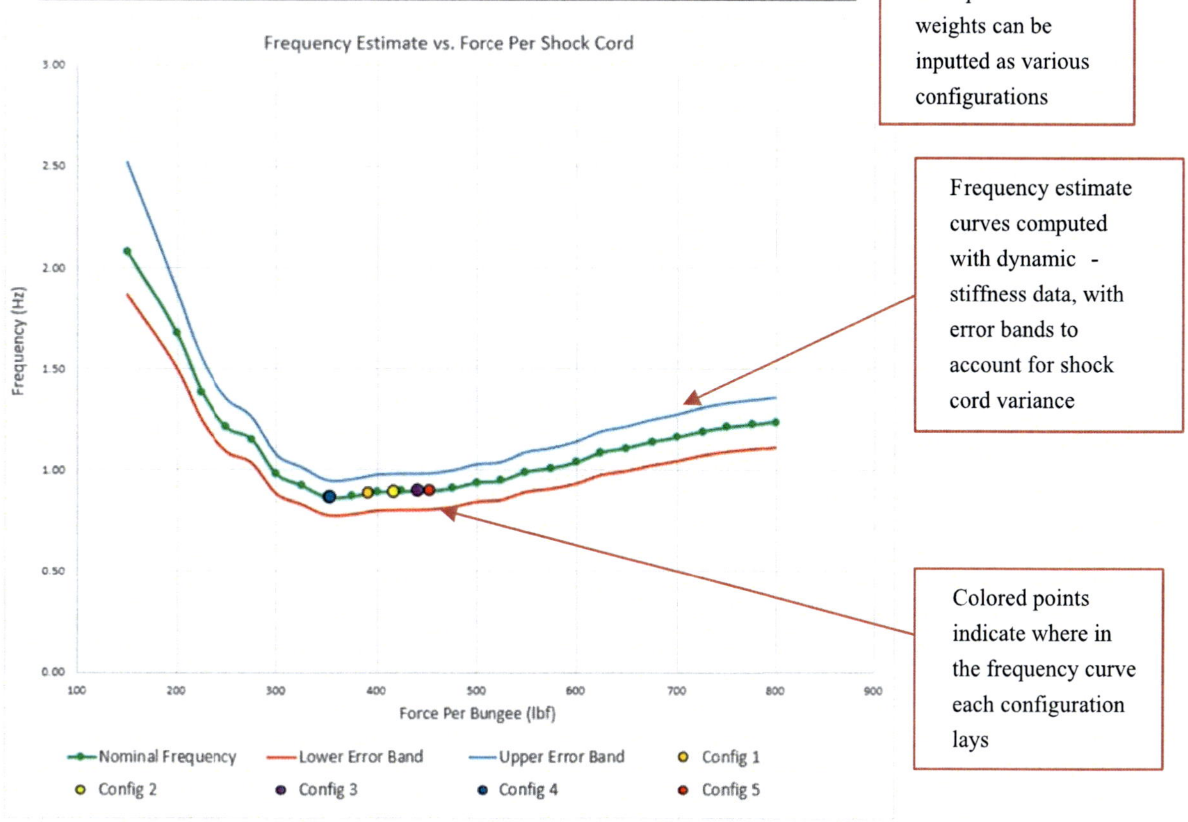

Fig. 14.6 Sample bounce mode estimator spreadsheet using dynamic-stiffness data

References

1. Brillhart, R.D., Napolitano, K.: Aircraft GVT Advances and Application – Gulfstream G650. In: Proceedings of the 29th International Modal Analysis Conference (February 2011)
2. Osterholt, D.J., Kelly, T.: Ground Vibration Testing of the World's Longest Wingspan Aircraft—Stratolaunch. In: Proceedings of the 29th International Modal Analysis Conference (February 2020)
3. Osterholt, D.J., Linehan, D.: Modal Test and Suspension Design for the Orion Launch Abort System. In: Proceedings of the 28th International Modal Analysis Conference (February 2010)
4. Whitney, M.J., Panza, J.L., Brillhart, R.D.: GVT Suspension Effects on Correlation. In: Proceedings of the 10th International Modal Analysis Conference (February 2002)

Chapter 15
Flight Worthiness Evaluation of Small Unmanned Aircraft Using Acoustic Testing

William Semke and Djedje-Kossu Zahui

Abstract A study on the ability to evaluate flight worthiness of a popular small unmanned aircraft systems (sUAS) using acoustic methods is presented. The ability to detect propeller and other potential damage that may be present in an autonomous operations environment is of great interest. SUAS are increasingly being designed to operate autonomously, however prior to flight there must be a flight worthiness evaluation to help ensure safe operations. To accomplish this, an evaluation of the effects of prop damage in the acoustic signature provides meaningful insight using a noninvasive evaluation tool. This study provides data obtained by experimentally measuring the acoustic noise levels produced, as well as the vibration levels on the host airframe and at the base. The data are analyzed to gain further understanding of the acoustic and vibration responses to accurately make predictions on flight worthiness. The experiments utilize an airframe that is commonly used in the UAS community with props with and without damage. The data are obtained from an aircraft powered in a stowed configuration, as would be occurring at a remote UAS launch station. This detailed study on the acoustic responses help enable the development of noninvasive damage detection algorithms. These systems would be part of robust and economical safety assessment procedures and protocols for preflight testing to ensure flight worthiness of remote autonomous sUAS operations that are planned.

Keywords UAS · Damage detection · Flight worthiness · Acoustic

15.1 Introduction

As small unmanned aircraft systems (sUAS), or drones, become ever more prevalent in commercial uses, new ways to utilize these devises are being proposed. One such area is the deployment of remote sUAS to conduct surveillance or monitoring activities in a remote autonomous fashion. These sites would be far from any personnel and must be able to be operated from great distances. While the communication links are available, one major factor that must be considered is the preflight inspections that occur prior to flight operations. This is typically done by a visual inspection, and then certain features are exercised prior to flight to make sure all components are functioning properly. In this work, appropriate remote inspection enhancements using acoustic measurements are proposed to better assess flight worthiness prior to mission operations in a robust and economical manner.

Many researchers have investigated the vibration environment on UAS. Sas et al. [1] introduce experimental and numerical analysis of a multirotor aircraft to find areas of lesser vibrations. Their goal was to identify places of lesser vibrations so they could mount the more sensitive electronics in those areas. Kloet et al. [2] identify the ambient noise of a multirotor UAS in two planes, adjacent to the propellers and under the propellers. Iannace et al. [3] offer a fault diagnosis method for UAVs using artificial neural networks. The model was trained to recognize the acoustics of the imbalanced propeller, to a success rate of 97% in the environment in which they tested. Previous work by Semke [4] offers analysis of different types of mounting mechanisms onboard a quadcopter and the vibration characteristics of the sensor mounting mechanisms of both multirotor and fixed wing aircrafts. The vibration environment onboard fixed-wing and quadrotor sUAS is presented in later work and provides sensor data to assist in passive and active vibration control methodologies [5]. In a previous paper by the authors, both vibration amplitudes of the aircraft body and acoustic measurements were made of an aircraft in flight [6]. Others have also compared propellers and the resulting vibration on a quadcopter [7]. Recent research was conducted by

W. Semke (✉) · D.-K. Zahui
Department of Mechanical Engineering, College of Engineering and Mines, University of North Dakota, Grand Forks, ND, USA
e-mail: william.semke@und.edu

C. Walber et al. (eds.), *Sensors & Instrumentation and Aircraft/Aerospace Testing Techniques, Volume 8*, Conference Proceedings of the Society for Experimental Mechanics Series, https://doi.org/10.1007/978-3-031-34938-6_15

Hu et al. [8] on the correlation between motor vibration and UAV stability. This study experimentally measures vibration levels and analyzed the correlation quantitatively. Li et al. [9] have designed a module to reduce the vibration observed in a payload from the excitation generated by the UAV. Many other studies have been done to provide vibration reduction in payload systems [10, 11].

Related work on wind turbine blades is pertinent in this study due to the many similarities of the UAS prop. Du et al. provide a comprehensive review of state-of-the-art damage detection techniques for wind turbine blades, including methods based on strain measurement, acoustic emission, ultrasound, vibration, thermography, and machine vision [12]. Inalpolat and Niezrecki provide a general overview of the different aspects of a novel structural health monitoring technique developed for wind turbine blades. Their proposed technique is based upon detecting the changes in acoustic transmission loss due to structural deformation and damage incurred [13].

15.2 Experimental Testing

Testing was performed on a DJI Phantom 4 Pro sUAS in a securely stowed configuration, as shown in Fig. 15.1, as well as simply setting on the tabletop without additional constraint. The accelerometer is either mounted to the bottom of the airframe directly below the center of the aircraft on the mounting bracket or the tabletop, depending on the test being conducted. The microphone is placed 10 cm directly above three test locations, which are above the center of the aircraft, above one of the props, or two rotor lengths past the edge of the aircraft. The tests were conducted using new DJI OEM 9450S Quick-Release Propellers and then again after a prop was damaged, as seen in Fig. 15.2. The aircraft was started and the testing was conducted while the props spun at the idle speed. The idle speed was chosen to maintain the speed across all the testing conducted. Higher speeds were investigated, but it was more difficult to recreate the testing environment and more noise was observed, thereby reducing the repeatability and reliability of the testing.

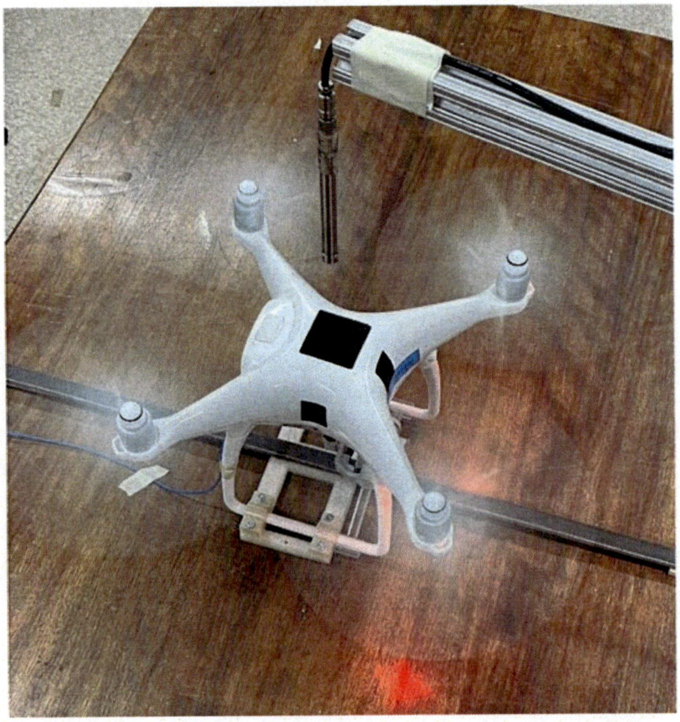

Fig. 15.1 DJI Phanton 4 Pro used for testing I the stowed configuration with the props spinning at idle speed

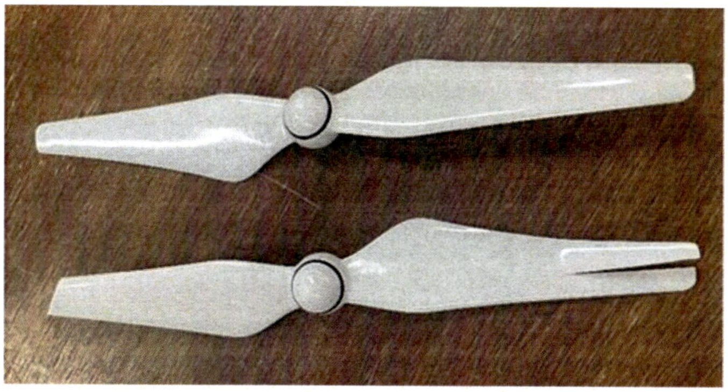

Fig. 15.2 DJI OEM 9450S Quick-Release Propellers, the upper is an undamaged prop and the lower is a damaged prop; broken tip on the left and split tip on the right

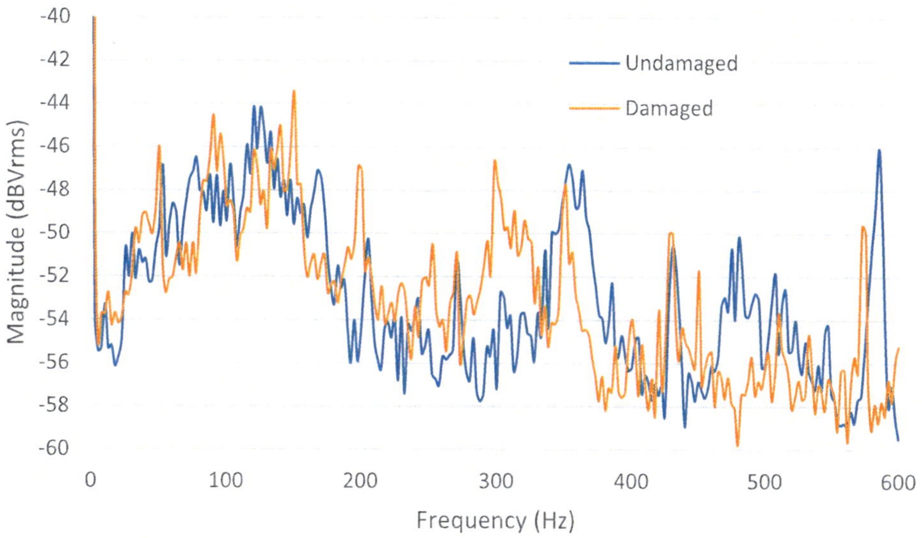

Fig. 15.3 Accelerometer FFT comparison of undamaged and damaged propellers measured onboard the aircraft with rigid mounting

15.3 Vibration Testing

The first set of experimental data presented is the accelerometer testing that was done. To measure vibrations, a PCB Model 333B30 single-axis accelerometer (10.5 mV/m/s^2) was attached to the base of the aircraft or directly below the center of the aircraft on the mounting bracket or on the table surface. The data was collected using the Dynamics Systems Analyzer (DSA) virtual instrument within the National Instruments myDAQ data acquisition system. The rotors were powered and kept at the idle speed during the test.

The first data sets were collected while the aircraft was secure in a storage configuration holding the landing gear firmly in place against the tabletop support. The frequency response function for the testing is shown in Figs. 15.3 and 15.4 for the accelerometer data collected onboard the aircraft and at the mounting base for the secure rigid mounting cases. The data in Fig. 15.3 do not show significant change between the damaged and undamaged cases. The biggest change observed can be seen in the increased contribution in frequencies around the 300 Hz and 500 Hz ranges. The authors do not believe the data obtained from the aircraft are reliable and, therefore, not definitive damage detection scheme. In Fig. 15.4, where the accelerometer is attached to the mount directly underneath the aircraft, a much greater impact resulting from the damage is observed. While much of this graph does not show substantial change, there is major change in the FFT magnitude in the 50 Hz range. There is clearly a significant impact on the response that it easily observed and may be used to indicate prop damage.

The second data sets were collected while the aircraft was simply sitting on the tabletop without any additional constraints. The frequency response function for the testing is shown in Figs. 15.5 and 15.6 for the accelerometer data collected onboard the aircraft and on the tabletop directly below the center of the aircraft. The data is Fig. 15.5 do not show significant change

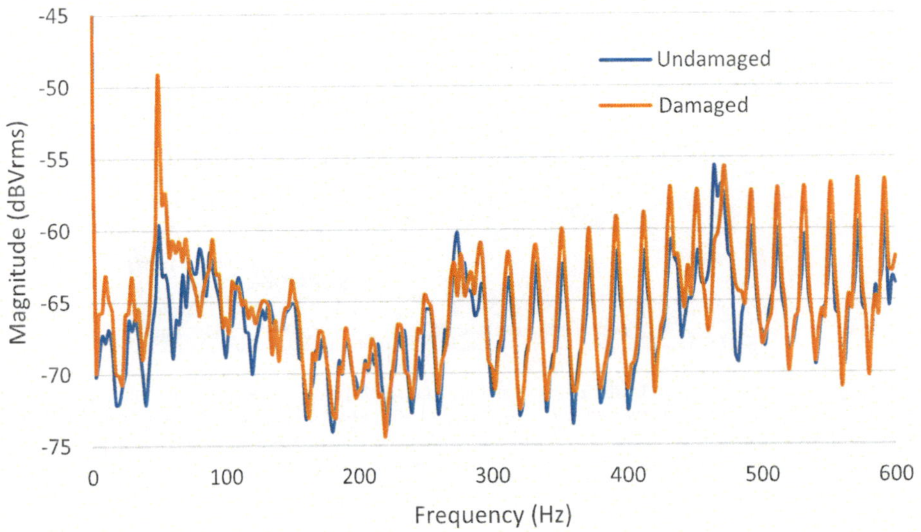

Fig. 15.4 Accelerometer FFT comparison of undamaged and damaged propellers measured at the mounting base with rigid mounting

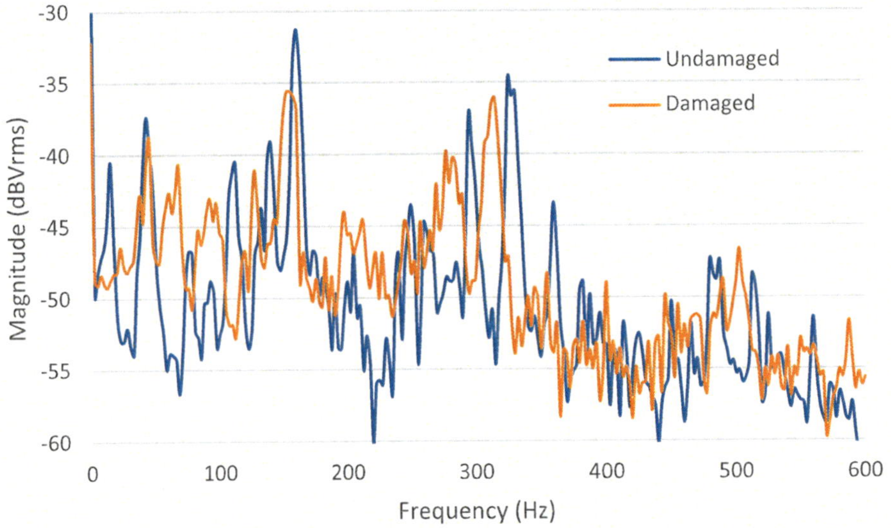

Fig. 15.5 Accelerometer FFT comparison of undamaged and damaged propellers measured onboard the aircraft without additional constraints

between the damaged and undamaged cases from the airframe accelerometer, although there is some shifting of the peaks throughout the range shown. The authors do not believe the data obtained from the aircraft is a reliable and definitive damage detection scheme. In Fig. 15.6, where the accelerometer is mounted on the tabletop underneath the center of the airframe, again a much greater impact resulting from the damage is observed. A major change in the FFT magnitude in the 50 Hz range is clearly seen, which is a significant change that it easily observed and may be used to indicate prop damage.

15.4 Acoustic Testing

To measure noise, the ACO Pacific 4012 – 1/2 Inch directional microphone with a PS9200 power supply was used. The raw data were recorded using a TEAC LX-10 and later imported into MALAB® for analysis. Using a MATLAB® power spectrum analyzer object, a voltage spectrum was generated. The spectrum was then converted into Sound Pressure Lever (SPL) using the microphone sensitivity (50 mV/Pa). The results were converted into sound level[3] at each frequency in Pascal (Pa) and subsequently in dB with a 20 μPa reference pressure using Eq. (15.1).

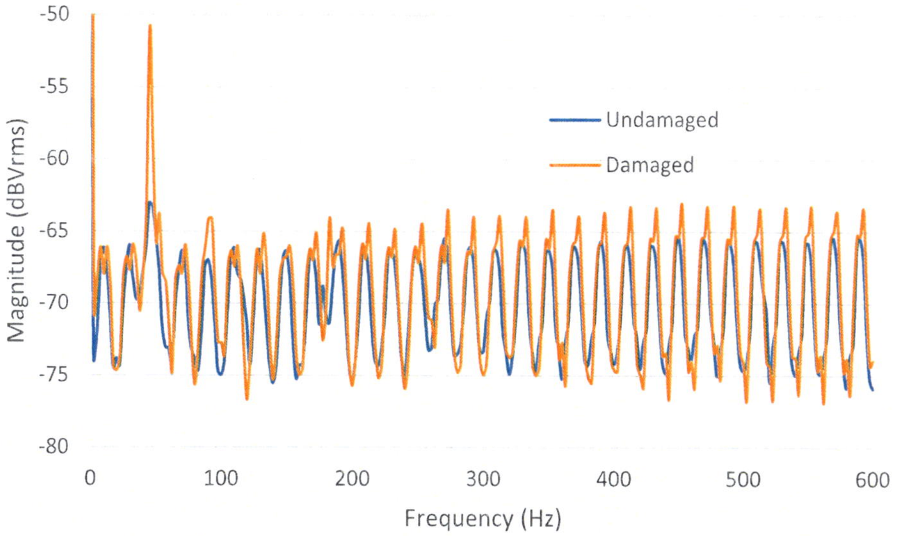

Fig. 15.6 Accelerometer FFT comparison of undamaged and damaged propellers measured at the mounting base without additional constraints

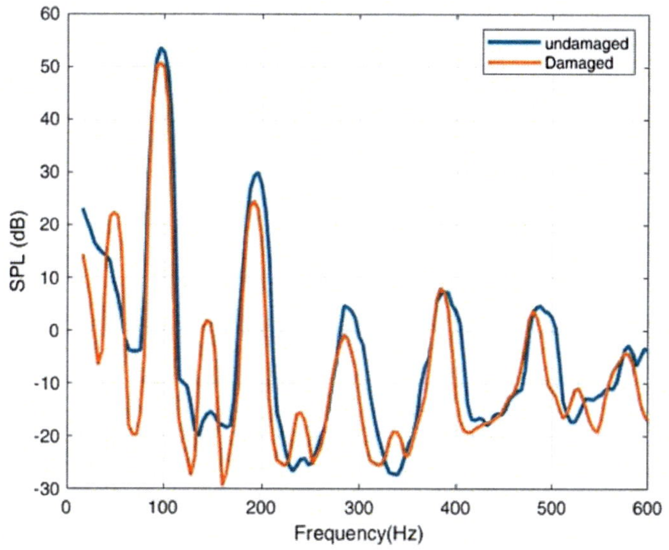

Fig. 15.7 Sound pressure spectrum comparison of damaged and undamaged propellers with the microphone placed 10 cm above the center of the aircraft with rigid mounting

$$L_i = 20\log_{10}\left(p_i/p_{\text{ref}}\right) \tag{15.1}$$

where L_i is the sound pressure in dB of the ith frequency and p_i sound pressure in Pa.

Figures 15.7, 15.8, and 15.9 depict the noise level spectrum for the three tests, where the microphone is above the center of the aircraft, above a rotor, and two blade lengths away while the aircraft is at idle speed and is rigidly mounted to the support structure. The three noise level spectrums shown in the figures follow the same basic response, although there are some differences as well. The measurements above the center of the aircraft have the strongest signal, due to proximity to all four rotors. The least response is when the microphone is two rotor lengths away from the aircraft. Therefore, the data indicate that a single microphone above the center of the aircraft is the best option as it provides a strong signal that can equally detect damage to any of the four rotors. The microphone above the rotors does not clearly produce more informative data; therefore, the use of four separate microphones, one above each rotor, is not warranted. The acoustic testing also showed a major change in the FFT magnitude in the 50 Hz range, which is a significant indicator that it easily observed and may be used to indicate prop damage.

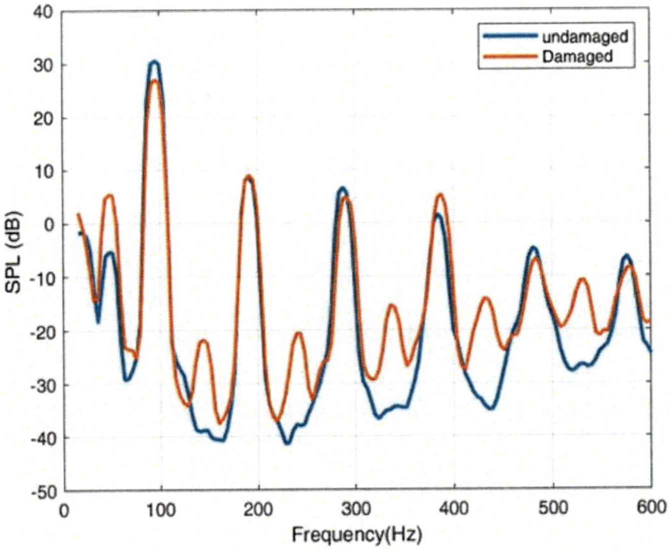

Fig. 15.8 Sound pressure spectrum comparison of damaged and undamaged propellers with the microphone placed 10 cm above one of the rotors of the aircraft with rigid mounting

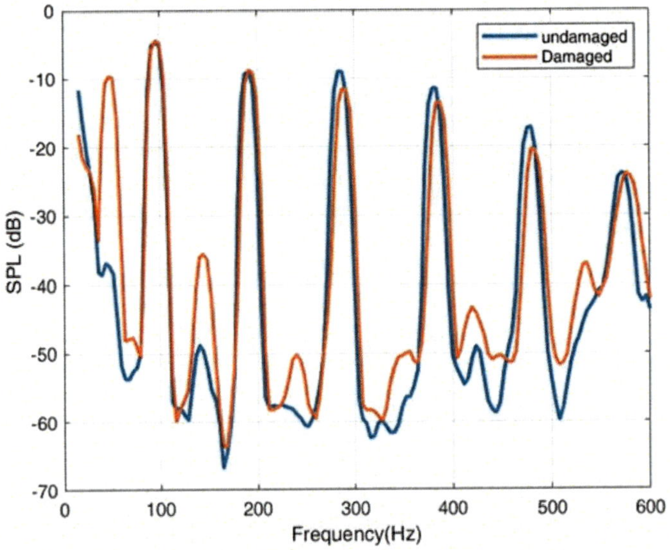

Fig. 15.9 Sound pressure spectrum comparison of damaged and undamaged propellers with the microphone placed two blade lengths from one of the rotors with rigid mounting

Figures 15.10 depicts the noise level spectrum for the case where the microphone is above the center of the aircraft while the aircraft is at idle speed and is sitting on the support structure without any additional restraints. The acoustic testing in this configuration also showed a major change in the FFT magnitude in the 50 Hz range and closely resembles the response in the secure mounting configuration.

15.5 Discussion

An additional study was carried out to evaluate the use of an inexpensive USB style microphone to replace the laboratory grade microphone. This was done to evaluate its effectiveness as potentially a means to utilize a web camera and microphone as a combination sensor for both the acoustic testing and visual inspection tool. The results are shown in Fig. 15.11. The figure shows the frequency up to 120 Hz to better focus on the primary area of interest in identifying damage. As can be

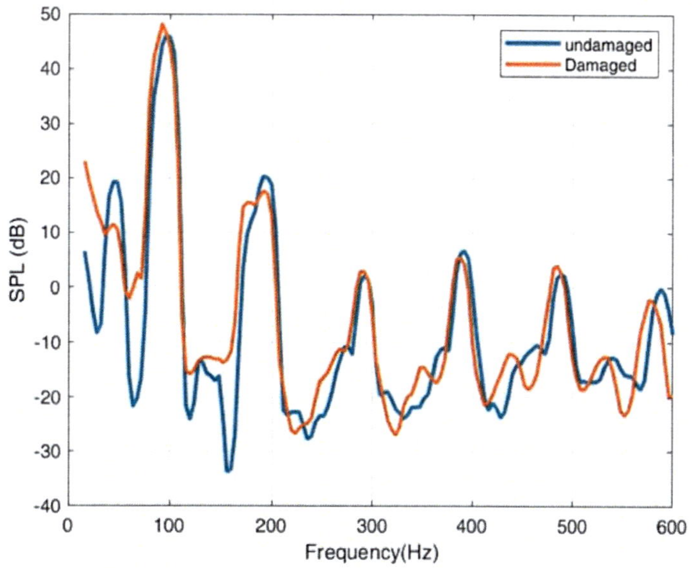

Fig. 15.10 Sound pressure spectrum comparison of damaged and undamaged propellers with the microphone placed above the center of the aircraft without additional constraints

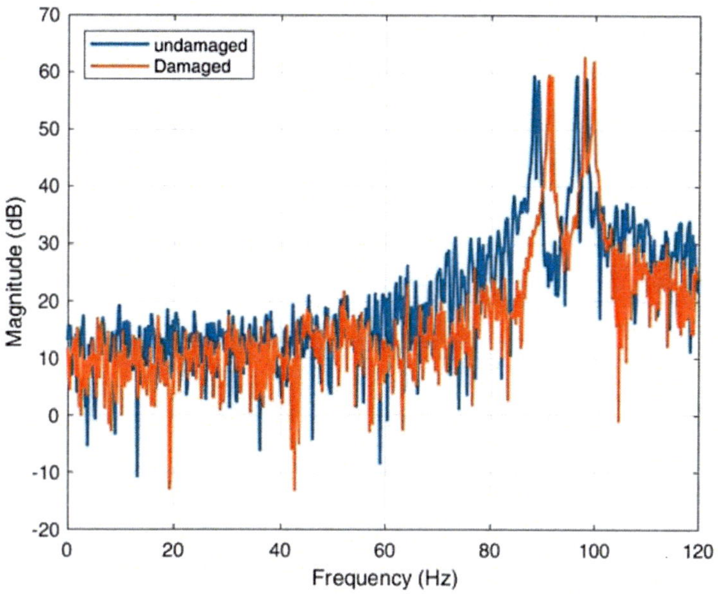

Fig. 15.11 Sound pressure spectrum comparison of damaged and undamaged propellers with the USB-style microphone placed above the center of the aircraft without additional constraints

seen, there is no indication of any change in this region of interest in the corresponding FFT. The inexpensive microphone is not responsive in this frequency range and therefore does not provide informative data to act upon. Only one USB-based microphone was assessed, and there may be other models or types that provide a more appropriate response.

In the testing presented, there were clear changes in the acoustic response observed. The changes to the props and/or other failure modes are manifested in the acoustic signature of the testing. A period of learning the characteristics and flight worthiness correlation will be required to understand the phenomena that are expected to be observed. This learning process could be done using a neural network or other artificial intelligence algorithm to more fully automate the process.

The use of acoustic methods over other methods is preferred due to the relatively low cost and ease of implementation. The required hardware is small and has low power consumption, making it appropriate for remote operations with limited upkeep and maintenance. This method aids in the deployment of remote sUAS to conduct surveillance or monitoring activities that are desired. These sites would be far from any personnel and have to be able to be operated from great distances. While the communication links are generally available, one major factor that must be considered is the preflight inspections that occur prior to flight operations. This is typically done by a visual inspection, and then certain features are exercised prior to flight to make sure all components are functioning properly. Now it has been shown there is reason to believe this can be effectively and efficiently done at a distance while maintaining a high level of safety.

15.6 Conclusion

The ability to evaluate flight worthiness of a sUAS using acoustic methods has been shown to be feasible. The ability to detect propeller and other potential damage using an acoustic sensor is meaningful, since prior to flight, there must be a flight worthiness evaluation to help ensure safe operations. The proposed noninvasive evaluation tool used on a common airframe shows the changes in the acoustic signature that could be correlated to prop damage. This detailed study helps enable the development of more comprehensive damage detection algorithms. These systems could be part of robust and economical safety assessment procedures and protocols for preflight testing to ensure flight worthiness of remote autonomous sUAS operations that are planned. The correlation of the frequency response to the extent of the damage and subsequent flight worthiness is still in need of development. Other inspection tools recommended are digital cameras to visually inspect the aircraft.

While the exact requirements for beyond visual line of site operations (BVLOS) of sUAS is still being developed by the Federal Aviation Administration (FAA), the concept of unmanned outposts with sUAS will require certain flight readiness procedures to be in place to ensure safe flight operations. It is expected that visual inspections will be required; however, this proposed acoustic interrogation will provide much more detailed information and could reveal unseen conditions that could cause flight performance issues that could result in crashes. In addition, the same procedure could be used to evaluate all sUAS in a preflight check to identify potential issues and serve as a health monitoring system, resulting in more effective scheduled maintenance.

References

1. Sas, P., et al.: Vibration Analysis of a UAV Multirotor Frame. KU Leuven - Department Werktuigkunde, Vibration Analysis of a UAV Multirotor Frame (2016)
2. Kloet, N., et al.: Acoustic signature measurement of small multi-rotor unmanned aircraft systems, pp. 3–14. Int. J. Micro Air Vehic. (2017). https://doi.org/10.1177/1756829316681868
3. Iannace, G., Ciaburro, G., Trematerra, A.: Fault diagnosis for UAV blades using artificial neural network. Robotics. **8**, 59 (2019)
4. Semke, W., Vibration reduction for camera systems onboard small unmanned aircraft. In: Proceedings of the International Modal Analysis Conference (IMAC) XXXVII: A Conference and Exposition on Structural Dynamics (2019)
5. Semke, W., Dunlevy, M.: A Review of the Vibration Environment Onboard Small Unmanned Aircraft. In: Proceedings of the International Modal Analysis Conference (IMAC) XXXVI: A Conference and Exposition on Structural Dynamics (2018)
6. Semke, W., Zahui, D., Schwalb, J.: The Vibration and Acoustic Effects of Prop Design and Unbalance on Small Unmanned Aircraft. In: Proceedings of the International Modal Analysis Conference (IMAC) XXXVII: A Conference and Exposition on Structural Dynamics (2020)
7. He, M., et al.: Comparison of Propellers and Reduction of Vibration in Four-Rotor Unmanned Aerial Vehicle (UAV). In: et al. Proceedings of IncoME-V & CEPE Net-2020. IncoME-V 2020. Mechanisms and Machine Science, vol 105. Springer, Cham (2021). https://doi.org/10.1007/978-3-030-75793-9_48
8. Experimental Research on UAV Stability Based on Vibration Signal Measurement of UAV Motors. In: 2021 3rd International Symposium on Robotics & Intelligent Manufacturing Technology (ISRIMT) (2021), pp. 131–134. https://doi.org/10.1109/ISRIMT53730.2021.9596915
9. Li, J., Lei, Z., Gao, D., Xu, K., Shi, G.: Design and Research on UAV Vibration Resistance Module. In: 2021 7th Annual International Conference on Network and Information Systems for Computers (ICNISC) (2021), pp. 550–553. https://doi.org/10.1109/ICNISC54316.2021.00104
10. Changshuai, Y.: Vibration test and vibration reduction design of UAV load radar. In: Proceedings of the 2019 4th International Conference on Automation, Control and Robotics Engineering. [S.l.]:: Association for Computing Machinery (2019)
11. Li, C., Tan, F.: Effect of UAV vibration on imaging quality of binary optical elements. In: 2018 IEEE International Conference on Mechatronics and Automation (ICMA) (2018), pp. 1693–1698. https://doi.org/10.1109/ICMA.2018.8484289
12. Du, Y., et al.: Damage detection techniques for wind turbine blades: A review. Mech. Syst. Signal Process. **141** (2020)
13. Inalpolat, M., Niezrecki, C.: Acoustic sensing based operational monitoring of wind turbine blades. J. Phys.: Conf. Ser. Bristol. **1452**(1) (2020). https://doi.org/10.1088/1742-6596/1452/1/012050

Chapter 16
Low-Order Mechanical Modeling of Liquid Fuel Sloshing

Morgan Choi and Huinam Rhee

Abstract Since the space launch vehicle undergoes not only external disturbances such as aerodynamic force or solar wind pressure but also internal dynamic influences such as the bending mode vibration of the vehicle and the sloshing of the liquid propellant, it is essential to reflect these disturbances when designing a precise flight controller. The sloshing phenomenon refers to the vibration of a fluid having a free surface and occurs in the liquid fuel and oxidant tanks due to the accelerated motion of the vehicle. Since the propellant's sloshing natural frequencies may overlap with the frequency band of the attitude control device. For this reason, the sloshing mode of the propellant may resonate with the actuation of the TVC (thrust vector control) device or the bending mode of the vehicle, which can cause severe damage of the vehicle or instability of the attitude control (Healy, Development of the Rocket Engine for the Jupiter Missile. Rocketdyne, 1958). Among the solutions to mitigate the sloshing problem, there is a method of installing baffles that provides damping in the sloshing, but in the case of a space vehicle, the baffle design causes an increase in mass and cost, so there is a limit to its use. Therefore, it is necessary to accurately capture the dynamic characteristics of the sloshing and to apply the low-order equivalent mechanical model to the attitude controller using a simple mechanical mass-spring model or pendulum model as shown in Fig. 16.1.

Keywords Sloshing · Equivalent mechanical model · Computational fluid dynamics · Multibody dynamics · KSLOSH

This chapter deals with the equivalent low-order mechanical modeling method of propellant sloshing for an arbitrary axisymmetric shape tank, and the results were verified through computational fluid analysis and multibody dynamics analysis. There is no analytical solution for the sloshing mode of liquid propellant in an axisymmetric shape fuel tank except for a perfect cylinder shape [2]. In this study, the mode shape and natural frequency of the propellant sloshing inside the axisymmetric shape tanks were numerically calculated using the Rayleigh-Ritz method and the principle of least action. The Lagrangian was derived in terms of the velocity potential, which was assumed to be a linear combination of the test functions through the Rayleigh-Ritz method. The equation derived through the principle of least action is expressed as a generalized eigenvalue problem, which determines the sloshing frequencies and corresponding mode shapes (Fig. 16.1).

The equivalent mechanical model of sloshing must have the same mass and center of mass as the actual propellant and produce the same force and torque. Equivalent mechanical models consist of one fixed mass and several dynamic masses and consist of a spring-mass or pendulum system. The force and torque generated by the sloshing of the propellant were calculated by integrating with respect to the propellant tank wall. The masses of the equivalent mechanical model were determined so that the sloshing force is the same as the actual sloshing. The locations of the masses were determined to produce the torque as the actual sloshing. Similar studies have been already conducted in numerous references (such as Dodge, The New Dynamic Behavior of Liquids in Moving Containers, Southwest Research Institute, 2000), but there exist some unclear contents in their details in public domains. In this study, an independent research for the sloshing was performed in detail, and consequently, a computer code KSLOSH was developed to generate the equivalent mechanical model of sloshing for an arbitrary axisymmetric tank. The verification of KSLOSH code was successfully performed by

M. Choi
Department of Aerospace Engineering, Sunchon National University, Suncheon, Jeollanam-do, Republic of Korea

H. Rhee (✉)
School of Mechanical and Aerospace Engineering/Center for Aerospace Research, Sunchon National University, Suncheon, Jeollanam-do, Republic of Korea
e-mail: hnrhee@sunchon.ac.kr

C. Walber et al. (eds.), *Sensors & Instrumentation and Aircraft/Aerospace Testing Techniques, Volume 8*, Conference Proceedings of the Society for Experimental Mechanics Series, https://doi.org/10.1007/978-3-031-34938-6_16

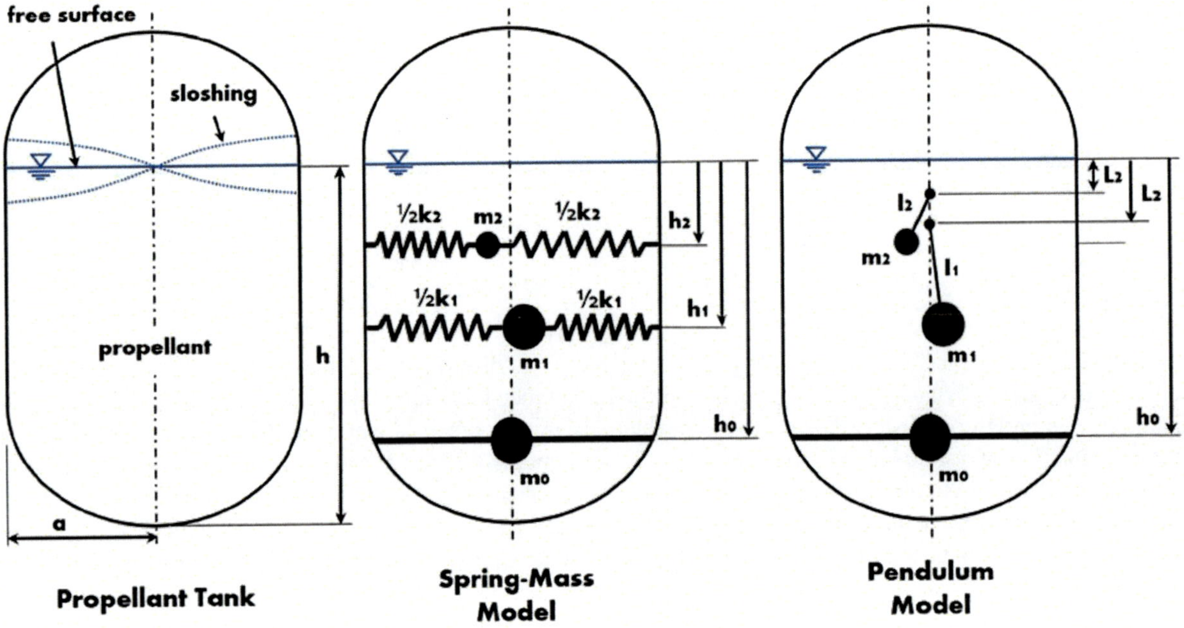

Fig. 16.1 Equivalent mechanical models of liquid propellant's sloshing

comparing the results from the computational fluid dynamics analysis and the multibody dynamic simulation of sloshing using the mechanical sloshing parameters determined by KSLOSH code.

Acknowledgements This study was supported by the National Research Foundation (NRF-2021R1I1A3059758).

References

1. Healy, R.: Development of the Rocket Engine for the Jupiter Missile. Rocketdyne (December 1958)
2. Dodge, F.T.: The New Dynamic Behavior of Liquids in Moving Containers. Southwest Research Institute (2000)

Chapter 17
Development of Steering Law for Thrust Vector Control Using Clustered Thrusters

Jiwoong Kim, Morgan Choi, and Huinam Rhee

Abstract Space launch vehicles often require large thrust engines depending on their mission. However, as the size of the engine increases, combustion instability problems may become severer and the reverse-thrust control, which is one of the key re-landing technologies required for the development of reusable rockets, becomes more difficult. Therefore, use of clustered low thrust engines in Fig. 17.1 has advantages compared to a single large thrust engine. In addition, use of the clustered engines makes it possible to control the yawing of the rocket, which is the spin motion about the longitudinal axis.

Keywords Thrust vector control · Electro-mechanical-actuator · Clustering · Steering law · Multibody dynamics

For a clustered engine, the number and arrangement of small thrust engines vary depending on the purpose of the vehicle. This may lead to an increase in the development cost as the steering law need to be changed. In this study, a generalized steering law for driving clustered rocket engines was developed through a kinematic approach. The proposed steering law can control the attitude of the space vehicle regardless of the number and arrangement of small thrust engines (Fig. 17.1).

The steering law of a single engine was first established. The TVC (Thrust Vector Control) engine controlled by linear motors consists of a linear motor connected by a ball joint and a universal joint. The engine thrust vector required to generate the required torque can be expressed as the product of the thrust vector based on the engine fixed coordinates and the rotation transformation matrix composed of the engine rotation angle [1]. The rotation transformation matrix is expressed as the product of two rotation transformation matrices as the engine rotates on two axes, and the order of the product of the rotation transformation matrices is determined by the axis of the universal joint. A universal joint is composed of a follower shaft and a base shaft; the axis to be rotated is determined as the follower shaft, and the axis connected to the fixed part is determined as the base shaft. In this chapter, the base shaft is determined as the axis where the engine mount and universal joint are connected and the follower shaft is determined as the axis where the engine and the universal joint are connected [2].

By determining the appropriate rotation, transformation matrix is applied to the joint position of the linear motor, the increment of the length of the linear motor is finally calculated through the position of the linear motor in the rotated coordinates, and the length of the linear motor calculated through the position of the linear motor. Through this process, when the required thrust is given to a single engine, the driving rule for calculating the increment of the linear motor length was established.

Then the torque generated by the clustered engine can be expressed as the product of the thrust vector of each engine and the position of the universal joint of each engine, and the thrust of the clustered engine can be expressed as the sum of the thrust of each engine. Therefore, the vector composed of the required torque and thrust can be expressed as the product of the required thrust vector of each engine, the matrix composed of the position of the universal joint of each engine, and the matrix composed of the unit matrices. This means that the required thrust vector of each engine can be calculated through matrix operation. Finally, it is possible to calculate the thrust vector of each engine through the pseudo-inverse and to determine the increment amount of the linear motor length of each engine through the driving rule of a single engine.

J. Kim · M. Choi
Department of Aerospace Engineering, Sunchon National University, Suncheon, Jeollanam-do, Republic of Korea

H. Rhee (✉)
School of Mechanical and Aerospace Engineering/Center for Aerospace Research, Sunchon National University, Suncheon, Jeollanam-do, Republic of Korea
e-mail: hnrhee@sunchon.ac.kr

C. Walber et al. (eds.), *Sensors & Instrumentation and Aircraft/Aerospace Testing Techniques, Volume 8*, Conference Proceedings of the Society for Experimental Mechanics Series, https://doi.org/10.1007/978-3-031-34938-6_17

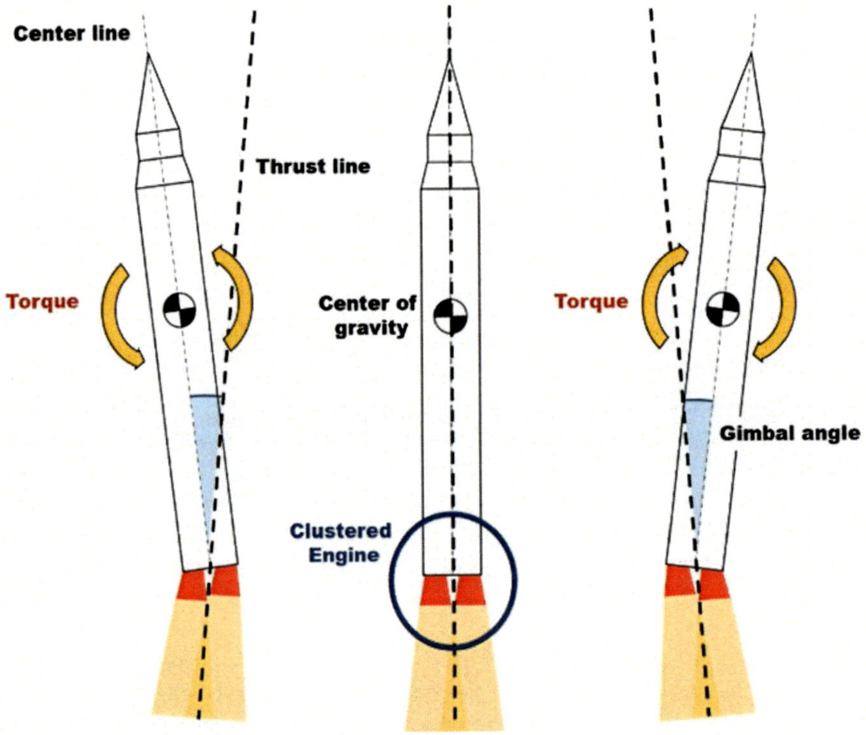

Fig. 17.1 Thrust vector control using clustered rocket engine

In order to verify the developed steering law, driving simulation was conducted using MATLAB Simulink and the multibody dynamics analysis program ADAMS. In this chapter, referring to the first-stage clustered engine of KSLV-2, recently developed in the Republic of Korea, four engines were modeled by clustering them in a circular arrangement [3]. The thrust and linear motor length of each engine calculated by the steering law using Simulink were input to the ADAMS, and then the force and torque generated by ADAMS were measured to check they exactly match with the required force and torque.

In this research, the TVC steering law applicable regardless of the number and arrangement of small engines for a clustered engine was successfully developed. The results can be used not only to reduce the development cost when developing space vehicles with various purposes but also as a basis for research when developing vehicles that require rapid maneuvering through nonlinearity analysis of the steering laws.

Acknowledgments This study was supported by the National Research Foundation of Korea (NRF-2021R1I1A3059758).

References

1. Li, Y.: Posture control of electromechanical-actuator-based thrust vector system for aircraft engine. NIEEE Trans. Indus. Electron. **59**, 3561 (2012)
2. Kim, J., et al.: Study on the EMA-based Thrust Vector Controller's Steering Law. In: SASE 2022 Spring Conference (May 2022)
3. Lee, H.J., Seung Hyub, O.: KSLV-II 1st stage clustered engine arrangement concept. Aerosp. Eng. Technol. (July 2012)

Chapter 18
Fiducial Marker–Based Localization of Autonomous UAV for Structural Health Monitoring

Ali Waqas and Young-Jin Cha

Abstract This chapter presents an autonomous navigation system based on computer vision for unmanned aerial vehicles (UAVs). The emerging area of autonomous UAV has shown rapid development in the field of structural health monitoring (SHM) for the collection of important structural data from civil infrastructure. Much of previous research has been focused on the use of magnetic compass sensor-based methods for control and localization of UAV in GPS-denied environments. However, these methods are vulnerable to magnetic interference of the surrounding environment and result in poor control of UAV with dangerous levels of path deviations. Therefore, this chapter suggests the use of unique small fiducial markers that are unsusceptible to magnetic interference and can be permanently attached to walls or ceilings of structures to enable robust autonomous flight of UAVs. A pseudo markers map is created where the location of each marker is stored in a marker's library. The markers are detected by the UAV during the flight and the prebuilt library information is used to localize UAV's position in the whole structure. It has been shown through various experiments that the use of computer vision-based technique can significantly improve the control of UAV in challenging indoor and outdoor GPS-denied environments. This marker-based localization method is a low-cost, flexible, and practical solution to realize true autonomous UAVs in civil structures such as bridges, parkade, and buildings.

Keywords UAV · Camera · SHM · Autonomous · Vision

18.1 Introduction

Unmanned aerial vehicles (UAVs) have gained increasing attention in the field of surveillance and inspection for collection of important data, that is, videos and images using the equipped visual sensors [1, 2]. UAVs offer many advantages over traditional methods, including lower costs, increased safety, and greater flexibility. UAVs can be used in areas which are unsafe and difficult to access due to its maneuverability and accurate control. Image and video data are of prime importance in structural health monitoring because it can be used to detect various surface damages. In this regard, one study showed the detection of cracks in RGB images using light-weight deep convolutional neural network [3]. In another research, unsupervised deep learning algorithms were used to detect damage [4, 5]. Unsupervised damage detection is important because it doesn't require creation of labeled dataset, which is a time-consuming work. Moreover, UAVs can also be equipped with other sensors such as thermal or infrared to acquire relevant data critical to specific inspection and monitoring purposes [6, 7]. The infrared and thermal images can be used to detect subsurface damage in bridges [8, 9], tunnels [10], and parkade [11]. Moreover, RGB images have also been used to quantify damage by estimating its volume [12]. In this study, deep learning model Faster R-CNN was used to detect concrete spalling and volumetric analysis were conducted to quantify the damage.

For this purpose, remote control–based UAVs have been used to get aerial photography and images of structures for monitoring purposes [13]. The remote control method cannot be carried out frequently due to its dependency on the availability of trained UAV pilots. As a result, researchers developed autonomous navigation systems using the global positioning system (GPS) sensors. The GPS sensors can be used to control a UAV autonomously over a predefined waypoint in outdoor environment with a clear view of the sky. However, in structures such as bridges, construction sites, and buildings, GPS signals are blocked by the presence of concrete and steel structures [14].

A. Waqas · Y.-J. Cha (✉)
Department of Civil Engineering, University of Manitoba, Winnipeg, MB, Canada
e-mail: young.cha@umanitoba.ca

C. Walber et al. (eds.), *Sensors & Instrumentation and Aircraft/Aerospace Testing Techniques, Volume 8*, Conference Proceedings of the Society for Experimental Mechanics Series, https://doi.org/10.1007/978-3-031-34938-6_18

To overcome the limitations of GPS in GPS-denied environments, four different methods have been proposed by researchers which can be categorized into (1) electromagnetic waves–based; (2) microelectromechanical systems (MEMS) inertial measuring unit (IMU)–based; (3) simultaneous localization and mapping (SLAM)–based and (4) artificial markers–based approaches. Electromagnetic waves–based techniques use radio frequency and ultrasonic signals to localize UAV in a GPS-denied environment. Kang et al. proposed the use of ultrasonic beacon system (UBS) for navigation of UAV in indoor environment to detect surface cracks in building infrastructure [15]. The proposed method consists of four stationary beacons and one mobile beacon. The mobile beacon can be localized by measuring its distance from each stationary beacon. This approach provides real-time accurate localization data in the form of spatial coordinates of UAV. The technique was further expanded to monitor cracks, loosened bolts, and corrosion in a steel bridge [14]. However, it requires that the sensors have a line of sight, and in real-world bridges with columns and beams, this is not practical to achieve. Moreover, they used magnetic compass sensors for estimation of UAV direction, which is unreliable in environments with high magnetic interference [16].

IMU-based approaches utilize the accelerometer and a gyroscope to estimate the yaw, pitch, and roll of UAV [17]. IMU is used due to its high sampling rate and low cost of installation, but the measurements are drifted over time and error accumulates since there is no ground truth mechanism to correct the error over time [18]. Like UBS, the IMU methods rely on magnetic sensors for yaw calculation, making them vulnerable to magnetic interferences due to steel structures.

SLAM is another approach to navigation and localization of UAV in GPS-denied environments which uses the features in real-world structures to construct a real-time map and localize the UAV within the map. ORB-SLAM [19] provides excellent robust performance in areas with a lot of visual features. However, in areas with simple walls and ceilings, SLAM performance degrades and loses tracking. Aguilar et al. [20] combined visual depth camera to implement SLAM, thus relying on the 3D data point cloud instead of only features in the structures. However, the range of 3D depth camera is only few meters and is not applicable in open areas, such as under the bridge.

Keeping in mind the challenges of UAV localization in civil infrastructure, fiducial markers [21] are utilized in our proposed approach. Bacik et al. proposed fiducial marker–based localization techniques [22]. Our proposed method extends the current fiducial marker–based approach even further to construct a virtual markers map which can be used for high precision maneuvering of UAV. The proposed method is depended only on computer vision techniques. These markers are easy to install on any surface, such as bridge columns or underneath the deck. The markers are detected by the UAV using the feed from camera and can be used to calculate the precise position and heading direction of UAV, as discussed in detail in the next section. Further inspiration for this methodology is drawn from our earlier publication [23].

18.2 Methodology

Our proposed methodology uses ArUco markers [24] for creation of 3D virtual map of the space where the UAV must navigate. ArUco markers are square-shaped planer markers with equal-sized black and white cells. Each marker consists of $n \times n$ cells, where n is the number of rows and columns, as shown in Fig. 18.1.

Each marker has its own local coordinate system (x_M, y_M, z_M) and four corner points (c_1, c_2, c_3, c_4). The binary marker is bordered by black cells, and the inner matrix of cells contains the data. This data can be used to detect the marker unique identification code or the markers id. This marker id can be used to store important information about the marker, such as the markers' location and orientation. A virtual markers map is created by attaching markers at various positions in the field and storing the location and orientation of each marker in a markers map.

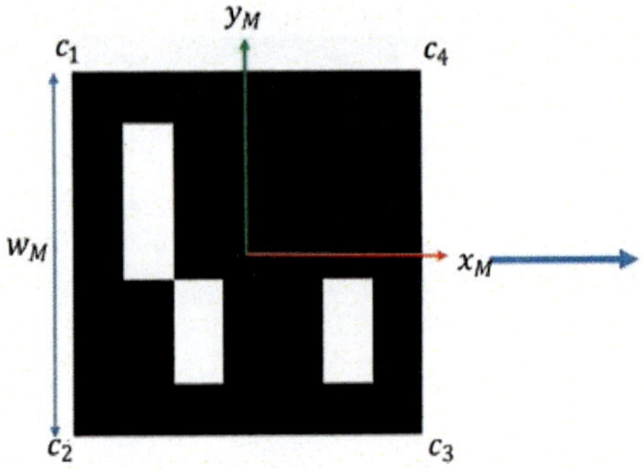

0	0	0	0	0	0	0
0	1	0	0	0	0	0
0	1	0	0	0	0	0
0	1	0	0	0	0	0
0	0	1	0	0	1	0
0	0	1	0	0	1	0
0	0	0	0	0	0	0

Fig. 18.1 Fiducial 7x7 ArUco markers with tag id 10 and its corresponding unique binary code

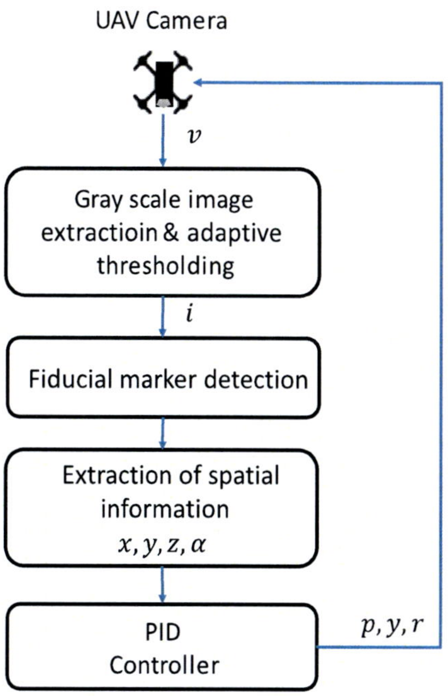

Fig. 18.2 Aruco marker–based localization of UAV

Next, the video stream v generated by UAV's camera is converted to grayscale image, as shown in Fig. 18.2. Then adaptive image thresholding is done to extract the most prominent edges and boundaries in the image frame i. Using the image i, fiducial markers is detected in the image using the method proposed by Romero-Romirez et al. [25]. After marker detection, the marker binary sequence code is extracted using Otsu's method [26].

Once marker is detected and code is extracted, it can be used to get the markers predefined position P_M in the map which was stored during installation time. Next, the orientation of UAV with respect to marker is calculated in the form of rotation matrix R. To calculate the rotation of marker with respect to UAV, the problem is treated as Perspective-n-point [27] problem from marker coordinate system to camera coordinate system. Here the position of corner points is known in marker and image coordinate system. Similarly, the intrinsic camera matrix A is also known using zhang's method [28]. Next, to calculate UAV spatial position coordinates (x_u, y_u, z_u) and yaw angle α we utilize the pinhole camera projection model as detailed in Eq. (18.1).

$$\begin{bmatrix} u \\ v \\ w \end{bmatrix} = A\,[R|T] \bullet C_i, \tag{18.1}$$

where u, v are pixel coordinates of corner point in image i and A is a 3×3 camera intrinsic matrix. R and T are the rotation and translation matrices of marker.

$$\begin{bmatrix} x \\ y \\ z \end{bmatrix} = R \bullet T + P_M, \tag{18.2}$$

where M is the position of installed marker. The required yaw angle α can be calculated directly from the rotation matrix R. Resultantly, it has been shown that localization information (x, y, z, α) can be calculated directly from the camera video feed v alone to generate the control values pitch, yaw, and roll (p, y, r) by the PID controller at ground station.

18.2.1 Outdoor Experiments

In this section, evaluation of our proposed localization and navigation is performed using outdoor experiments in real-world environment. The experiments were conducted at the University of Manitoba parkade on 400 level, as shown in Fig. 18.3, with Aruco marker encircled in red.

The UAV used in our experiments is the parrot Anafi [29] drone, which is a compact 320 g UAV equipped with high resolution 4 K camera and provides reliable Wi-Fi connectivity for data communication, as shown in Fig. 18.4.

The parkade had a lot of steel and concrete structures, due to which the GPS signals were weak and magnetic compass sensors were unreliable. We conducted various experiments in which the UAV was tasked to fly in a rectangular path of 3×15 meters, as shown in Fig. 18.5.

As shown in Fig. 18.4, the UAV was able to maneuver successfully using only the video feed from the UAV camera.

Fig. 18.3 Outdoor experiment setting with fiducial markers attached to the ceiling for UAV localization

Fig. 18.4 Images of UAV used in experiments

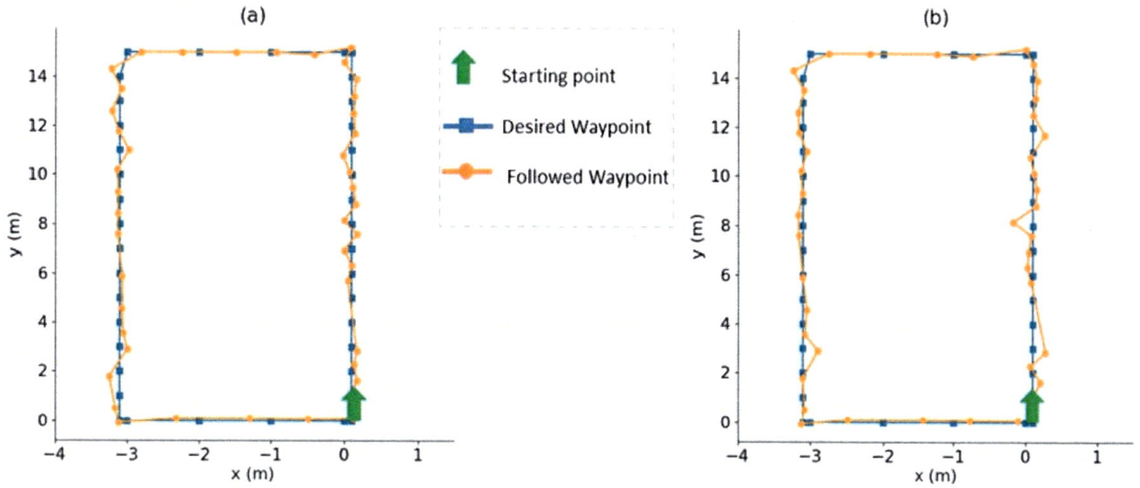

Fig. 18.5 Path followed by the UAV in outdoor experiment

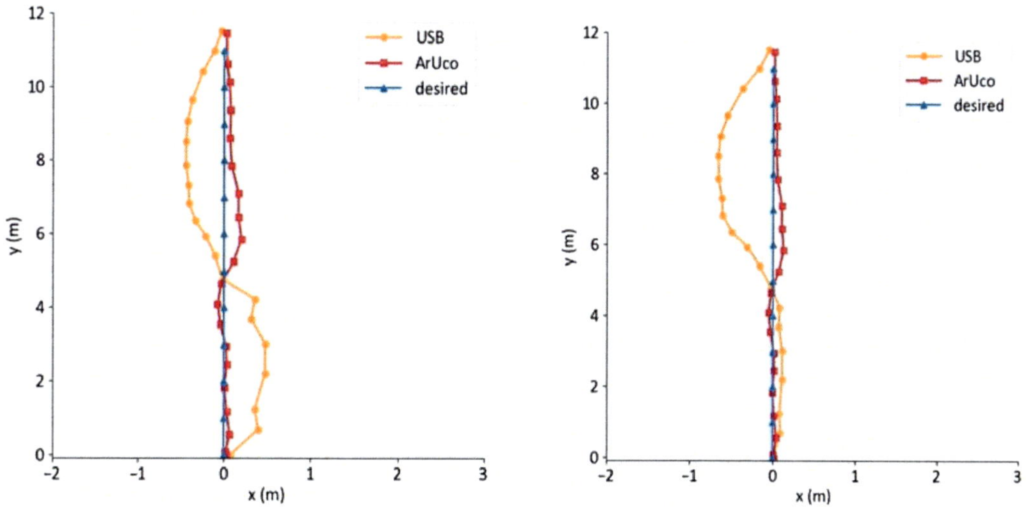

Fig. 18.6 Deviation of path of UAV using magnetic sensor–based method compared to fiducial marker–based method in magnetic interference and GPS-denied environment

Table 18.1 Comparative analysis of Magnetic sensor-based method vs ArUco marker–based localization

S. no.	Evaluation metric	Magnetic sensor	ArUco marker
1	Mean square error (MSE)	0.38	*0.058*
2	Standard deviation (SD)	0.30	*0.044*
3	Max deflection	0.48	*0.17 m*

Furthermore, we conducted various autonomous UAV experiments using both the fiducial marker–based and magnetic sensors–based yaw control experiments to quantify the improvement of our method in high magnetic interference environment, as shown in Fig. 18.6.

The results of the proposed fiducial marker–based method and magnetic interference methods are compared in terms of mean square error, standard deviation, and maximum path deflection in Table 18.1. The absolute path deviation is 0.48 m when using compass-based yaw control method as compared to 0.17 m when using fiducial marker–based methods.

18.3 Conclusion

This chapter presents a cost-effective alternative to localization and navigation of UAVs in GPS-denied environment using fiducial ArUco markers. The proposed method relies only on the use of camera video feed and is independent of magnetic sensors, which are impacted by the magnetic interference in the civil infrastructure. Moreover, this method has been shown to be effective, robust, and flexible to expand to larger areas. It requires one-time installation of special markers and can be applied in most of the civil infrastructure buildings.

Acknowledgments The research presented in this chapter was supported by the Canada Foundation for Innovation Grant (CFI JELF Grant No. 37394).

References

1. Nahangi, M., Heins, A., McCabe, B., Schoellig, A.: Automated localization of UAVs in GPS-denied indoor construction environments using fiducial markers. In: ISARC Proceedings of the International Symposium on Automation and Robotics in Construction (Vol. 35, pp. 1–7). IAARC Publications (2018)
2. Gonzalez, L.F., Montes, G.A., Puig, E., Johnson, S., Mengersen, K., Gaston, K.J.: Unmanned aerial vehicles (UAVs) and artificial intelligence revolutionizing wildlife monitoring and conservation. Sensors. **16**(1), 97 (2016)
3. Cha, Y.J., Choi, W., Büyüköztürk, O.: Deep learning-based crack damage detection using convolutional neural networks. Comput. Aided Civ. Inf. Eng. **32**(5), 361–378 (2017)
4. Wang, Z., Cha, Y.J.: Unsupervised deep learning approach using a deep auto-encoder with a one-class support vector machine to detect damage. Struct. Health Monit. **20**(1), 406–425 (2021)
5. Wang, Z., Cha, Y.J.: Unsupervised machine and deep learning methods for structural damage detection: a comparative study. Eng. Reports, e12551 (2022)
6. Israel, M., Reinhard, A.: Detecting nests of lapwing birds with the aid of a small unmanned aerial vehicle with thermal camera. In: 2017 International Conference on Unmanned Aircraft Systems (ICUAS), pp. 1199–1207. IEEE (2017, June)
7. Burke, C., Rashman, M., Wich, S., Symons, A., Theron, C., Longmore, S.: Optimizing observing strategies for monitoring animals using drone-mounted thermal infrared cameras. Int. J. Remote Sens. **40**(2), 439–467 (2019)
8. Ali, R., Cha, Y.J.: Subsurface damage detection of a steel bridge using deep learning and uncooled micro-bolometer. Constr. Build. Mater. **226**, 376–387 (2019)
9. Ali, R.: Deep learning-and infrared thermography-based subsurface damage detection in a steel bridge (Master's thesis) (2019)
10. Ali, R., Zeng, J., Cha, Y.J.: Deep learning-based crack detection in a concrete tunnel structure using multispectral dynamic imaging. In: Smart Structures and NDE for Industry 4.0, Smart Cities, and Energy Systems, vol. 11382, pp. 12–19. SPIE (2020, April)
11. Ali, R., Cha, Y.J.: Attention-based generative adversarial network with internal damage segmentation using thermography. Autom. Constr. **141**, 104412 (2022)
12. Beckman, G.H., Polyzois, D., Cha, Y.J.: Deep learning-based automatic volumetric damage quantification using depth camera. Autom. Constr. **99**, 114–124 (2019)
13. Benz, C., Debus, P., Ha, H.K., Rodehorst, V.: Crack segmentation on UAS-based imagery using transfer learning. In: 2019 International Conference on Image and Vision Computing New Zealand (IVCNZ), pp. 1–6. IEEE (2019, December)
14. Ali, R., Kang, D., Suh, G., Cha, Y.J.: Real-time multiple damage mapping using autonomous UAV and deep faster region-based neural networks for GPS-denied structures. Autom. Constr. **130**, 103831 (2021)
15. Kang, D., Cha, Y.J.: Autonomous UAVs for structural health monitoring using deep learning and an ultrasonic beacon system with geo-tagging. Comput. Aided Civ. Inf. Eng. **33**(10), 885–902 (2018)
16. Afzal, M.H., Renaudin, V., Lachapelle, G.: Assessment of indoor magnetic field anomalies using multiple magnetometers. In: Proceedings of the 23rd International Technical Meeting of The Satellite Division of the Institute of Navigation (ION GNSS 2010) (pp. 525–533) (2010, September)
17. Ibrahima, M., Moselhib, O.: IMU-based indoor localization for construction applications. In: ISARC Proceedings of the International Symposium on Automation and Robotics in Construction (Vol. 32, p. 1). IAARC Publications (2015)
18. Balamurugan, G., Valarmathi, J., Naidu, V.P.S.: Survey on UAV navigation in GPS denied environments. In: 2016 International Conference on Signal Processing, Communication, Power and Embedded System (SCOPES), pp. 198–204. IEEE (2016, October)
19. Mur-Artal, R., Montiel, J.M.M., Tardos, J.D.: ORB-SLAM: a versatile and accurate monocular SLAM system. IEEE Trans. Robot. **31**(5), 1147–1163 (2015)
20. Aguilar, W.G., Rodríguez, G.A., Álvarez, L., Sandoval, S., Quisaguano, F., Limaico, A.: Visual SLAM with a RGB-D camera on a quadrotor UAV using on-board processing. In: International Work-Conference on Artificial Neural Networks, pp. 596–606. Springer, Cham (2017, June)
21. Sani, M.F., Karimian, G.: Automatic navigation and landing of an indoor AR. drone quadrotor using ArUco marker and inertial sensors. In: 2017 International Conference on Computer and Drone Applications (IConDA), pp. 102–107. IEEE (2017, November)
22. Bacik, J., Durovsky, F., Fedor, P., Perdukova, D.: Autonomous flying with quadrocopter using fuzzy control and ArUco markers. Intell. Serv. Robot. **10**(3), 185–194 (2017)
23. Waqas, A., Kang, D., Cha, Y. J.: Deep learning-based obstacle-avoiding autonomous UAVs with fiducial marker-based localization for structural health monitoring. Struct. Health Monit., 14759217231177314

24. Garrido-Jurado, S., Muñoz-Salinas, R., Madrid-Cuevas, F.J., Marín-Jiménez, M.J.: Automatic generation and detection of highly reliable fiducial markers under occlusion. Pattern Recogn. **47**(6), 2280–2292 (2014). https://doi.org/10.1016/j.patcog.2014.01.005
25. Romero-Ramirez, F.J., Muñoz-Salinas, R., Medina-Carnicer, R.: Speeded up detection of squared fiducial markers. Image Vis. Comput. **76**, 38–47 (2018)
26. Otsu, N.: A threshold selection method from gray-level histograms. IEEE Trans. Syst. Man Cybern. **9**(1), 62–66 (1979)
27. Fischler, M. A., Bolles, R. C.: Random sample consensus: a paradigm for model fitting with applications to image analysis and automated cartography. Commun. of the ACM **24**(6), 381–395 (1981)
28. Zhang, Z.: A flexible new technique for camera calibration. IEEE Trans. Pattern Anal. Mach. Intell. **22**(11), 1330–1334 (2000)
29. Ackerman, E.: Parrot's new drone reclaims aniche: the Anafi marks the company's return to theconsumerspace-[Resources_Review]. IEEE Spectr. **55**(9), 21–21 (2018)

Chapter 19
Obstacle Avoidance Method for Autonomous UAV for Structural Health Monitoring

Ali Waqas and Young-Jin Cha

Abstract This chapter presents an obstacle avoidance method (OAM) to realize an autonomous collision-avoiding unmanned aerial vehicle (UAV) for the collection of structural data from civil infrastructure. OAM is of high importance to avoid serious accidents during autonomous flights of UAVs for monitoring purposes. A collision must be avoided to avert the loss of human and financial loss. Therefore, this chapter provides a new unique real-time OAM that consists of four steps: obstacle detection, obstacle clustering, distance estimation, and generation of new waypoints. For obstacle detection, deep learning algorithm YOLOv3 is implemented, which can detect obstacles at 10 frames per second during the flight. Obstacles are detected in the form of bounding boxes in the image stream of the UAV. Next, if more than one obstacle is detected, the k-means clustering algorithm is used to group the obstacles based on their relative position and the nearest obstacle group is selected to be avoided first. Then the distance from the nearest obstacle to the UAV is estimated using monocular depth estimation and a pinhole camera model. If the obstacle is dangerously close to the UAV, it must be avoided. For this purpose, a new obstacle avoidance waypoint is generated based on the obstacle position and flight path of the UAV. Experimentally, it has been shown that our obstacle avoidance method has real-time and robust performance compared to existing state-of-the-art OAM methods.

Keywords Obstacles · UAV · SHM · Camera · Drone

19.1 Introduction

Recently unmanned aerial vehicles (UAVs) have gained a lot of interest in the field of structural health monitoring (SHM) [1–3]. Manual inspection of bridges and roads using UAV is time-consuming and requires expert pilots. Therefore, autonomous UAVs have gained increased importance, especially in the field of civil engineering for structural health monitoring [4, 5]. These UAVs use the onboard camera to record photos and videos of the surface to be inspected. Cha et al. showed that the images can be used to detect cracks at pixel level in the RGB images by utilizing convolution neural network and deep learning algorithms [6]. Furthermore, it has been shown that autonomous UAVs can be used to detect multiple type of damages simultaneously, such as cracks, corrosion, and steel delamination [7]. Unsupervised algorithms have also been used to extract damage in civil infrastructure using RGB images [8–10]. These types of methods are important because it skips the time-consuming data collection and data labelling step. It is also helpful when data set is not already available for the type of damage to be detected. Furthermore, UAVs can be equipped with thermal camera to detect sub surface damage in concrete structures. Ali & Cha used thermal images to detect delamination in tunnel [11], parkade structures [12], and steel bridge [13, 14]. Thermal camera provides temperature distribution of the surface, which can be used to detect any type of anomaly in temperature distribution over the surface [13]. Other interesting studies have used deep learning algorithms on image data to detect damages in important civil structures [15, 16].

UAVs provide a very reliable, cost-effective, and feasible way for data collection from places where humans cannot safely reach. UAV is also extremely safe to use as no human life is at risk during the flight. Moreover, UAVs can reliably and autonomously collect data required for specific inspection task. Therefore, UAVs enable SHM field engineers to do frequent monitoring of structures. However, to achieve a truly autonomous flight of UAV, the UAV must be able to navigate itself in the environment while avoiding any type of obstacles that might be present, such as furniture, columns, girders, etc. Obstacles

A. Waqas · Y.-J. Cha (✉)
Department of Civil Engineering, University of Manitoba, Winnipeg, MB, Canada
e-mail: young.cha@umanitoba.ca

C. Walber et al. (eds.), *Sensors & Instrumentation and Aircraft/Aerospace Testing Techniques, Volume 8*, Conference Proceedings of the Society for Experimental Mechanics Series, https://doi.org/10.1007/978-3-031-34938-6_19

must be avoided because it can lead to serious accidents and loss of expensive equipment. These autonomous UAVs must be equipped with robust obstacle avoidance algorithms for field use cases. In this regard, researchers proposed several methods. One method uses color to distinguish between the road and obstacle by using a single RGB camera equipped with mobile robot [17]. This method is very simple and cannot be used in challenging real-world environment where obstacle can be of different colors. In another study, ultrasonic sensors were used to detect obstacle in close range of mobile robot. The sensor can be mounted on the UAV to detect obstacle [18]. However, this requires modification and addition of new sensor, which is not available on the UAV.

Another approach is to use deep learning to detect obstacles using the images from UAV's camera. Most of the object detection networks consist of a backbone architecture for feature extraction, such as ResNet [19], ResNext [20], DenseNet [21], MobileNet [22], DarkNet [23]. A head is attached to the backbone to generate the bounding box of detected obstacle. Head can be one-staged or two-staged for detection. R-CNN [24] operates as a two-stage detector. In contrast, YOLO [25], Single Shot Detector (SSD) [26], and RetinaNet [27] function as single-stage detectors. Single stage detector is usually fast and light weight compared to multistaged detectors. Therefore, we selected the YOLOv3 as the detection algorithm for our proposed OAM. As most UAVs are already equipped with camera, this proposed OAM can be applied to a variety of available UAVs in the market. This methodology has been inspired by our previously published work in reference [28].

19.2 Methodology

Our obstacle avoidance method uses the UAV's camera to detect obstacle. For detection, we have adopted the use of YOLOV3, which is one of the most robust object detection algorithms based on deep learning and convolution neural network [29]. YOLOv3 is also fast enough to realize obstacle avoidance in real time because it can process images at 172 frames per second, which is more than enough for real-time obstacle detection. The detection of YOLOv3 results in bounding box, as shown in Eq. (19.1).

$$\text{box} = \{b_x, b_y, b_h, b_w, c_1, \ldots c_n\} \tag{19.1}$$

where b_x, b_y is the center of the bounding box and $c_1 \ldots c_n$ are the probability of object class. We modified the YOLOv3 architecture to detect the specific class of obstacles that were used during experiments, which in our case are chairs, as shown in Fig. 19.1. YOLOv3 uses Darknet as feature extractor and uses YOLO head to extract bounding boxes.

Once obstacle is detected using YOLOv3, the output is in the form of multiple bounding boxes depending on YOLOv3 detection. These boxes need to be filtered out and cleaned to minimize the incorrect boxes in the prediction results, as seen in Fig. 19.2. We filter out boxes that are too small to be considered an obstacle.

Next, if multiple bounding boxes are detected, the bounding boxes are clustered using K-means [30] clustering algorithm and visualized as single bounding box. If the obstacle is very close to the UAV, the OAM decides whether to move to left or right depending on the UAV position in the image frame, as shown in Eq. (19.2).

$$\text{OAM Direction} = \begin{cases} \text{right}: & b_x \geq \frac{W}{2} \\ \text{left}: & b_x < \frac{W}{2} \end{cases} \tag{19.2}$$

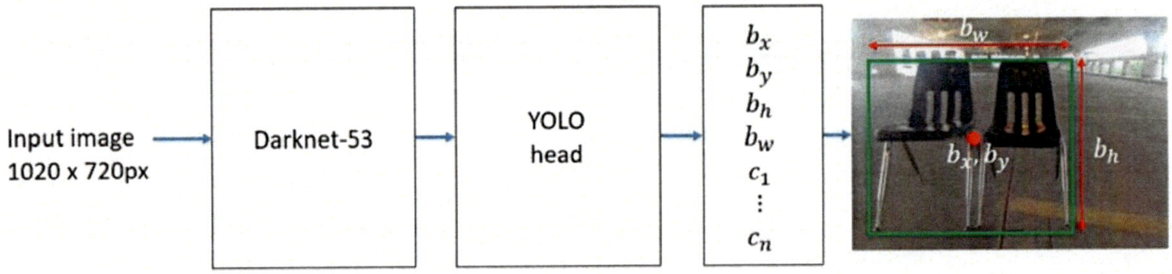

Fig. 19.1 Obstacle detection in image frame using YOLOv3

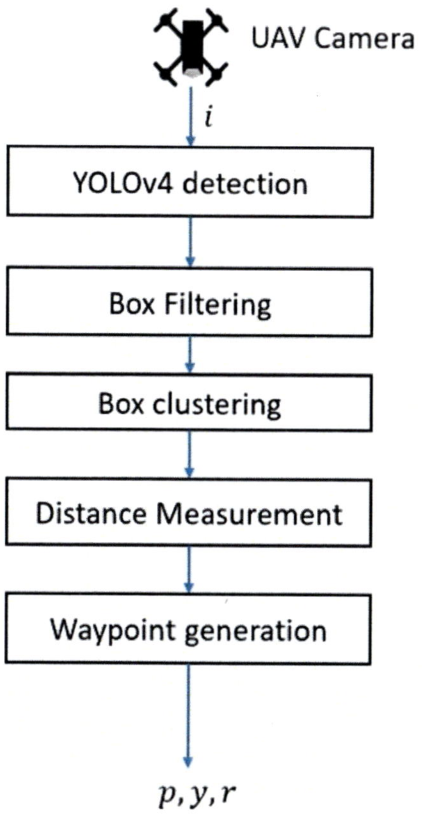

Fig. 19.2 Proposed obstacle avoidance method (OAM)

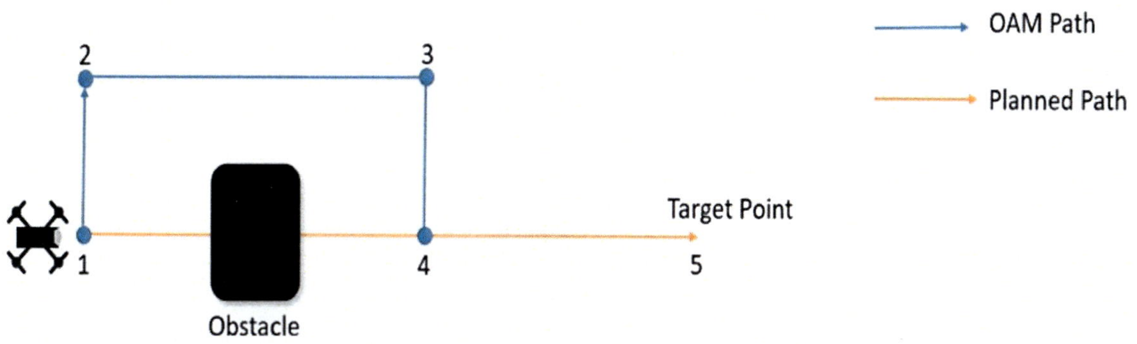

Fig. 19.3 OAM path planning during real-time flight

where W is the width of the image frame. Based on Eq. (19.2), the UAV moves left or right until it is safe to move straight and return to original planned path of the UAV, as shown in Fig. 19.3, from point 1 to 2. The OAM-generated path is $(1, 2, 3, 4, 5)$ as compared to original planned path $(1, 5)$.

19.3 Analysis

We performed various experiments of the method for testing purposes at the University of Manitoba parkade for validation of the proposed method. For experimental setup we used chairs as obstacle for the UAV. The chairs were placed in various configuration along the UAV planned trajectory to act as an obstacle as shown in Fig. 19.4. The threshold distance for avoidance was one meter and the image size to be processed by OAM was 1020 × 720 pixels. The Parrot Anafi [31] UAV, which was used for our experiments, is equipped with a high-quality full HD camera and Wi-Fi connectivity, allowing for image processing at the ground station. The camera is equipped with gimbal to provide stable photos with minute motion

Fig. 19.4 Configuration of obstacle in UAV path

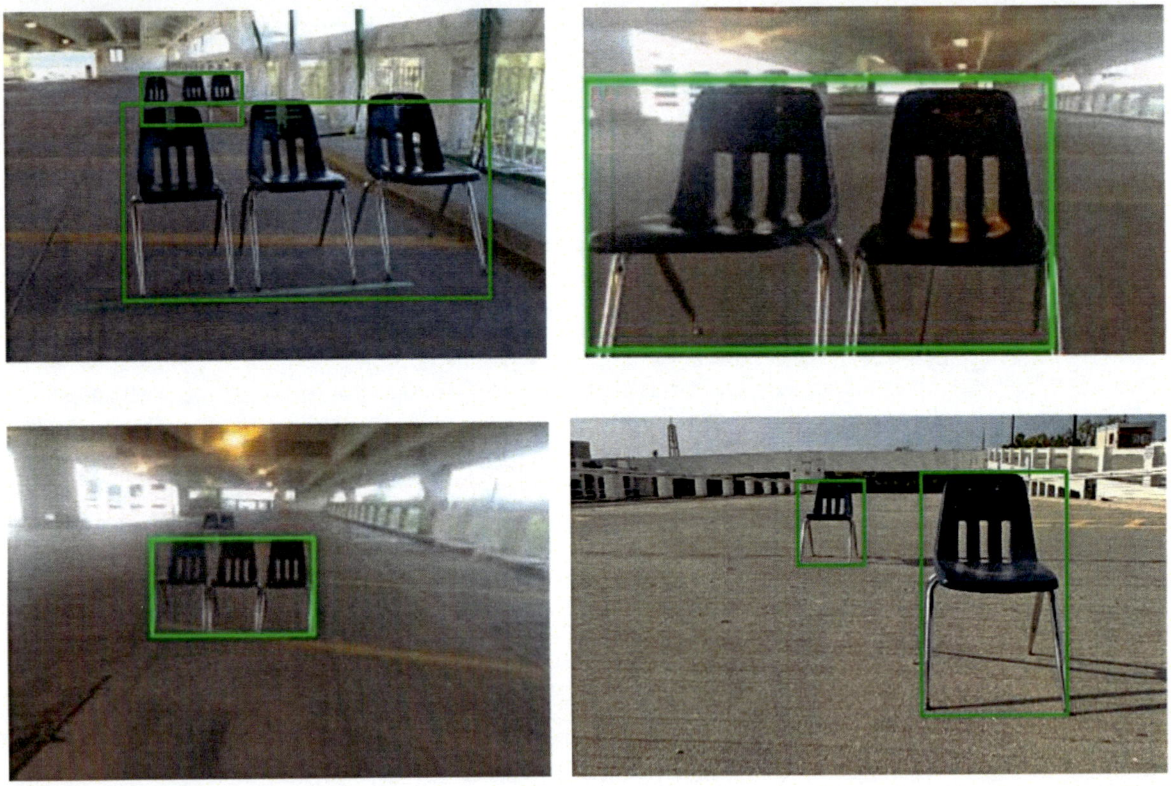

Fig. 19.5 Obstacle detection using YOLOv3

distortions. The UAV's video feed was continuously analyzed in real-time by the OAM, ensuring timely obstacle detection and decision-making at each waypoint throughout the flight.

The robust YOLOv3 algorithm was able to detect the obstacle in various configuration for the OAM to avoid the obstacle, as shown in Fig. 19.5. In the experimental setup, UAV was equipped with ultrasonic beacon system for localization and way finding. The OAM was integrated into the existing localization system. In this way, the proposed algorithm can be integrated into diverse range of localization modules for autonomous UAVs. The OAM was trained to detect chairs as obstacle, but it can be trained to detect multiple obstacles in real-world environment.

In this study, the UAV primary purpose was to collect data from the structure while simultaneously avoiding the obstacle in its path. In our experiments, it was seen that the UAV was able to avoid the obstacle placed at different points and in different configuration (Fig. 19.6).

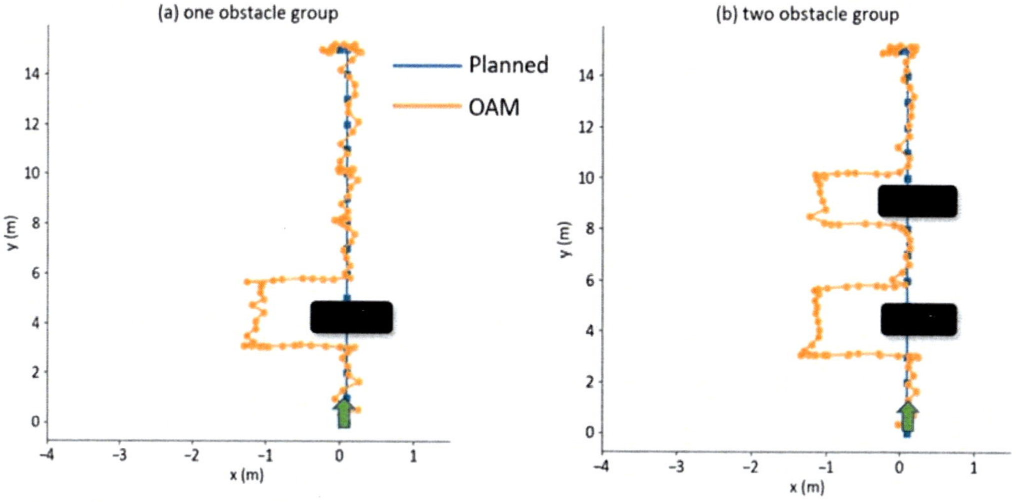

Fig. 19.6 Path followed by the UAV shown in orange compared to path planned before the flight using proposed OAM

19.4 Conclusion

This chapter proposes a new fast and robust obstacle avoidance method (OAM) for UAVs in real-world environment. The proposed OAM uses YOLOv3 to detect obstacles in the visual feed of the UAV and generates a new waypoint to avoid the obstacles. The proposed algorithm has been experimentally tested in a real-world structure at the University of Manitoba parkade, and its performance has been robust. The proposed OAM has the potential to be used in practical real-world environment by integrating with existing GPS-based or IMU-based localization modules.

Acknowledgments The research presented in this chapter was supported by the Canada Foundation for Innovation Grant (CFI JELF Grant No. 37394).

References

1. Kim, H., Lee, J., Ahn, E., Cho, S., Shin, M., Sim, S.H.: Concrete crack identification using a UAV incorporating hybrid image processing. Sensors. **17**(9), 2052 (2017)
2. Sankarasrinivasan, S., Balasubramanian, E., Karthik, K., Chandrasekar, U., Gupta, R.: Health monitoring of civil structures with integrated UAV and image processing system. Procedia Comp. Sci. **54**, 508–515 (2015)
3. Ayele, Y.Z., Aliyari, M., Griffiths, D., Droguett, E.L.: Automatic crack segmentation for UAV-assisted bridge inspection. Energies. **13**(23), 6250 (2020)
4. Kang, D., Cha, Y.J.: Autonomous UAVs for structural health monitoring using deep learning and an ultrasonic beacon system with geo-tagging. Comput. Aided Civ. Inf. Eng. **33**(10), 885–902 (2018)
5. Ali, R., Kang, D., Suh, G., Cha, Y.J.: Real-time multiple damage mapping using autonomous UAV and deep faster region-based neural networks for GPS-denied structures. Autom. Constr. **130**, 103831 (2021)
6. Cha, Y.J., Choi, W., Büyüköztürk, O.: Deep learning-based crack damage detection using convolutional neural networks. Comput. Aided Civ. Inf. Eng. **32**(5), 361–378 (2017)
7. Cha, Y.J., Choi, W., Suh, G., Mahmoudkhani, S., Büyüköztürk, O.: Autonomous structural visual inspection using region-based deep learning for detecting multiple damage types. Comput. Aided Civ. Inf. Eng. **33**(9), 731–747 (2018)
8. Amezquita-Sancheza, J.P., Valtierra-Rodriguez, M., Adeli, H.: Machine learning in structural engineering. Scientia Iranica. **27**(6), 2645–2656 (2020)
9. Wang, Z., Cha, Y.J.: Unsupervised deep learning approach using a deep auto-encoder with a one-class support vector machine to detect damage. Struct. Health Monit. **20**(1), 406–425 (2021)
10. Wang, Z., Cha, Y.J.: Unsupervised machine and deep learning methods for structural damage detection: a comparative study. Eng. Reports, e12551 (2022)
11. Ali, R., Zeng, J., Cha, Y.J. Deep learning-based crack detection in a concrete tunnel structure using multispectral dynamic imaging. In: Smart Structures and NDE for Industry 4.0, Smart Cities, and Energy Systems (Vol. 11382, pp. 12–19). SPIE (2020, April)
12. Ali, R., Cha, Y.J.: Attention-based generative adversarial network with internal damage segmentation using thermography. Autom. Constr. **141**, 104412 (2022)

13. Ali, R., Cha, Y.J.: Subsurface damage detection of a steel bridge using deep learning and uncooled micro-bolometer. Constr. Build. Mater. **226**, 376–387 (2019)
14. Ali, R.: Deep learning-and infrared thermography-based subsurface damage detection in a steel bridge (Master's thesis) (2019)
15. Kang, D.H., Cha, Y.J.: Efficient attention-based deep encoder and decoder for automatic crack segmentation. Structural Health Monitoring, 14759217211053776 (2021)
16. Choi, W., Cha, Y.J.: SDDNet: real-time crack segmentation. IEEE Trans. Ind. Electron. **67**(9), 8016–8025 (2019)
17. Ulrich, I., Nourbakhsh, I.: Appearance-based obstacle detection with monocular color vision. In: AAAI/IAAI (pp. 866–871) (2000, July)
18. Azeta, J., Bolu, C., Hinvi, D., & Abioye, A.A.: Obstacle detection using ultrasonic sensor for a mobile robot. In: IOP Conference Series: Materials Science and Engineering (Vol. 707, No. 1, p. 012012). IOP Publishing (2019, November)
19. He, K., Zhang, X., Ren, S., Sun, J.: Deep residual learning for image recognition. In: Proceedings of the IEEE Conference on Computer Vision and Pattern Recognition (pp. 770–778) (2016)
20. Morgenthal, G., Hallermann, N., Kersten, J., Taraben, J., Debus, P., Helmrich, M., Rodehorst, V.: Framework for automated UAS-based structural condition assessment of bridges. Autom. Constr. **97**, 77–95 (2019)
21. Huang, G., Liu, Z., Van Der Maaten, L., Weinberger, K.Q.: Densely connected convolutional networks. In: Proceedings of the IEEE Conference on Computer Vision and Pattern Recognition (pp. 4700–4708) (2017)
22. Howard, A.G., Zhu, M., Chen, B., Kalenichenko, D., Wang, W., Weyand, T., ... Adam, H.: Mobilenets: Efficient convolutional neural networks for mobile vision applications. arXiv preprint arXiv:1704.04861 (2017)
23. Redmon, J.: Darknet: Open source neural networks in c. http://pjreddie.com/darknet/ (2013–2016)
24. Girshick, R.: Fast r-cnn. In: Proceedings of the IEEE International Conference on Computer Vision (pp. 1440–1448) (2015)
25. Redmon, J., Divvala, S., Girshick, R., Farhadi, A. You only look once: unified, real-time object detection. In: Proceedings of the IEEE Conference on Computer Vision and Pattern Recognition (pp. 779–788) (2016)
26. Liu, W., Anguelov, D., Erhan, D., Szegedy, C., Reed, S., Fu, C.Y., Berg, A.C.: Ssd: single shot multibox detector. In: European Conference on Computer Vision (pp. 21–37). Springer, Cham (2016, October)
27. Lin, T.Y., Goyal, P., Girshick, R., He, K., Dollár, P.: Focal loss for dense object detection. In: Proceedings of the IEEE International Conference on Computer Vision (pp. 2980–2988) (2017)
28. Waqas, A., Kang, D., Cha, Y.J.: Deep learning-based obstacle-avoiding autonomous UAVs with fiducial marker-based localization for structural health monitoring. Structural Health Monitoring, 14759217231177314 (2023)
29. Bochkovskiy, A., Wang, C.Y., Liao, H.Y.M.: Yolov4: optimal speed and accuracy of object detection. arXiv preprint arXiv:2004.10934 (2020)
30. Hartigan, J.A., Wong, M.A.: Algorithm AS 136: a k-means clustering algorithm. J. R. Statist. Soc. Ser. C (applied statistics). **28**(1), 100–108 (1979)
31. Ackerman, E.: Parrot's new drone reclaims aniche: the Anafi marks the company's return to theconsumerspace-[Resources_Review]. IEEE Spectr. **55**(9), 21–21 (2018)

Chapter 20
Modal Characterization of 3D Printed Compliant Mechanisms for Space Exploration

Dorota Budzyń, Hossein Zare-Behtash, and Andrea Cammarano

Abstract The moon's surface is covered with dust (regolith) and microscopic particles created by numerous meteorite impacts (lunar meteoric gardening). The lack of the smoothening actions of hydrological and aeolian process and the interaction with cosmic radiation have made regolith abrasive and electrostatically charged. Due to these characteristics, regolith has been identified as a major issue to lunar (and other planets) exploration, being responsible for clogging and wearing in the exposed hinges of the tools used in extravehicular activities.

Compliant mechanisms permit to eliminate the use of hinges and, therefore, of sliding motions between adjacent surfaces, because they use elastic deformation to supply the desired kinematic behavior. Due to their design, these mechanisms present areas that undergo large deformations and alternating stress concentrations and, therefore, fatigue effects. In this work, we examine the dynamic behavior of a compliant mechanism designed for extravehicular space exploration using additive manufacturing. The vibration mode-shapes and the corresponding natural frequencies will be identified, with particular focus on the equivalent structural characteristics and their linearity.

Keywords Compliant mechanisms · Topology optimization · Dust mitigation

20.1 Introduction

Lunar dust, which is the finest fraction of lunar regolith, poses a challenge for surface exploration missions. During Apollo program, it had negative impact on the hardware used on the lunar surface [1–3]. The dust consists of sharp particles in micrometers range of sizes [4, 5]; such debris can easily enter the gaps in between the elements of the hardware. Furthermore, lunar soil consists of minerals that score at least 6 on the Mohs scale [5]. This is harder than most of the engineering materials used for scientific equipment. The dust is also electrostatically charged, increasing its adhesion to equipment surfaces. As the electrostatic behavior of dust changes with the day and night lunar cycle, it is also possible for the dust to temporarily float above the surface [6]. This can increase the range and speed of contamination.

There are multiple dust mitigation technologies, usually divided into active, passive, and implicit. Active methods use external forces to clean equipment from the dust. This could be as simple as using brushes, but other more complex methods like vibration, ultrasonic, or electrodynamic cleaning are under development [1, 7]. Passive methods do not require use of any external force, and they mitigate the dust attraction; they include work function matching coatings, microstructure films, etc. [1, 8]. The implicit dust mitigation technologies use design solutions that are naturally dust resilient. This work focuses on solving the problem of increased friction and jamming of rigid body mechanisms contaminated by the lunar dust. An implicit dust mitigation technology, in this case, should eliminate the mechanisms' inter-element gaps to avoid their penetration by the dust. A solution that can achieve this goal is to use novel mechanisms, namely, complaint mechanisms. There are multiple challenges related to the design and utilization of compliant mechanisms for use in space. As part of space missions, compliant mechanisms should be able to withstand the vibrations of launch systems. Furthermore, the modes of compliant mechanisms can be a limitation for their use in some applications, and therefore, it is necessary that the dynamic response of these mechanisms is fully characterized before they are used in extra vehicular activities. In addition, some of the characteristics of the dynamic response could provide the basis for further applications that have not been considered at the

D. Budzyń · H. Zare-Behtash · A. Cammarano (✉)
James Watt School of Engineering, University of Glasgow, Glasgow, UK
e-mail: andrea.cammarano@glasgow.ac.uk

C. Walber et al. (eds.), *Sensors & Instrumentation and Aircraft/Aerospace Testing Techniques, Volume 8*, Conference Proceedings of the Society for Experimental Mechanics Series, https://doi.org/10.1007/978-3-031-34938-6_20

time of their design. For example, although compliant mechanisms are more resilient to clogging, when covered in regolith, their thermal properties are affected, so other mechanisms may be considered to actively remove some regolith. Inducing electric travelling waves is one of the solutions currently considered for active removal. This can be coupled with structural vibrations, but this requires a dynamic characterization of the mechanisms. As such, dynamic simulations of compliant mechanisms are an interesting problem that can improve their design and future use in space. In this work, we will present a preliminary analysis of 3D printed compliant grippers designed using topology optimization and analytical methods.

20.2 Compliant Mechanisms

Compliant mechanisms use elastic deformation to achieve motion. They do not require multiple components and can be manufactured as monolithic pieces, which reduces the complexity of assembly [9]. The lack of inter-element gaps reduces tribological wear and avoids the need of lubrication, which is usually challenging in the space industry [10]. Furthermore, with the appropriate design, compliant mechanisms can be an excellent choice for precision mechanisms as they are free from backlashes [11–13]. Compliant mechanisms also store strain energy while deflected; when the input force is removed, compliant mechanisms can flex back to their original shapes acting like spring-loaded systems. Examples of compliant hinges that can be used to replace rigid-body hinges are presented in Fig. 20.1.

There are multiple methodologies for the design of compliant mechanisms. As compared to the design of rigid-body mechanisms, compliant mechanisms can be quite unintuitive to synthesize. Two main methodologies used to arrive at the compliant solution are analytical design methods and topology optimization. Analytical design methods use a library of compliant submechanisms that can form a more complex system, or they replace existing rigid-body joints with predefined compliant hinges with the same degrees of freedom. In this work, two grippers utilizing the Instant Centre Aproach will be presented. Figure 20.2 presents the gripper design (half of the design domain as the symmetry was assumed).

Topology optimization leaves the synthesis of compliant mechanism to software tools. Topology optimization is a methodology that searches for optimal material distribution for a given problem. In this work, we present outcomes of two topology optimization simulations aiming at achieving compliant gripper design. The simulation was set up using the displacement-driven formulation proposed by Koppen et al. [14]. The evolution of both topologies is visible in Fig. 20.2. The top row represents the optimization process that used PLA (polylactic acid) material parameters, and the bottom row shows the same optimization setup but with material changed to TPC (thermoplastic copolyester) (Fig. 20.3).

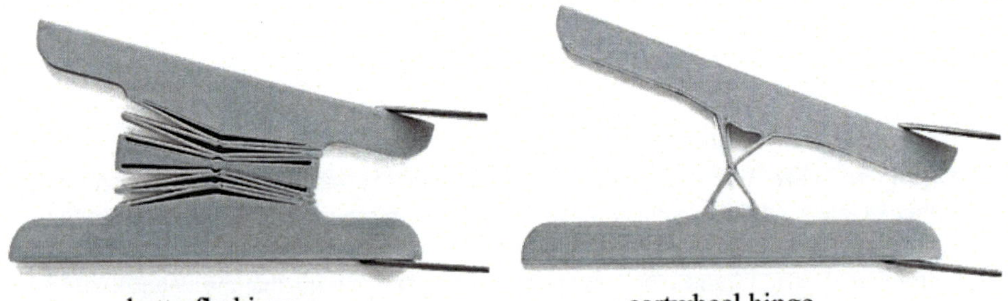

Fig. 20.1 Examples of compliant hinges

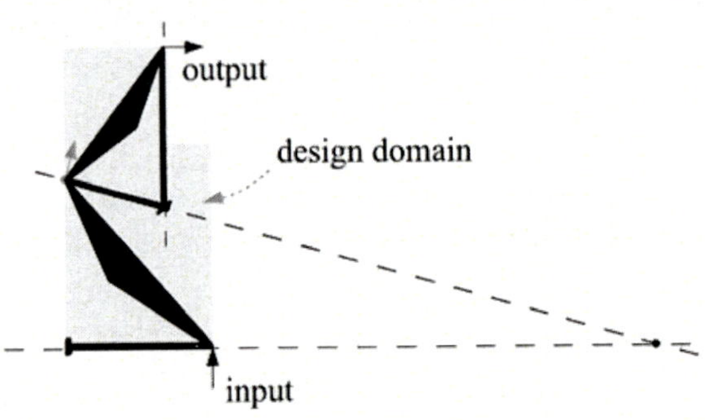

Fig. 20.2 Instant Center Approach design sketch of compliant gripper

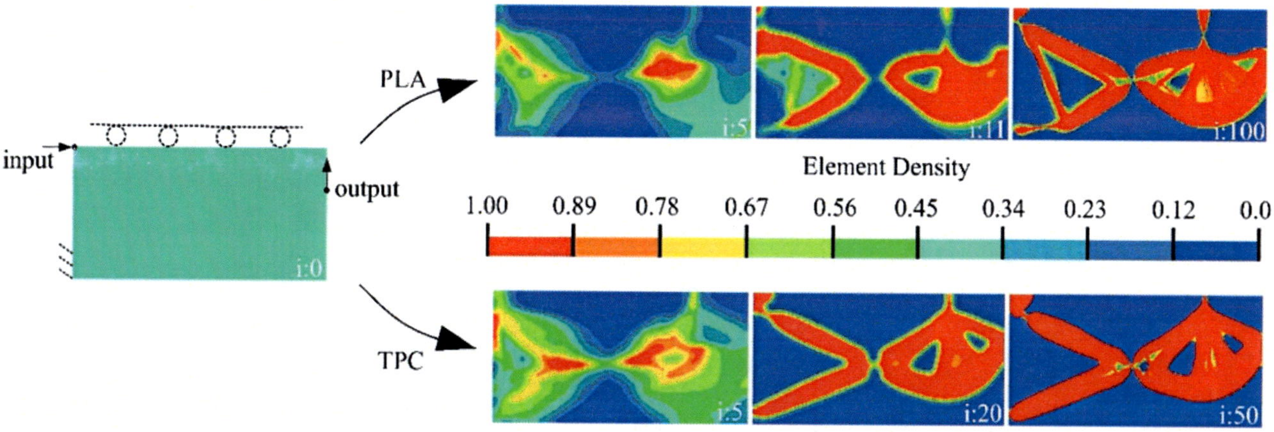

Fig. 20.3 Topology optimization design domain (left) and topology optimization progression for two different materials: PLA on the top and TPC on the bottom ('i' – number of current iteration)

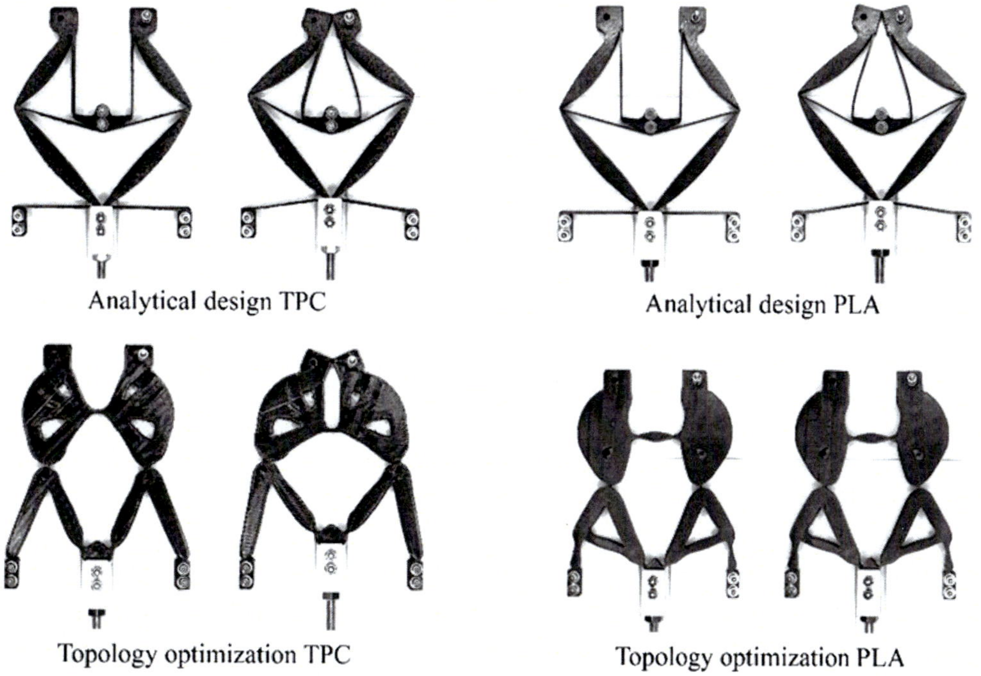

Fig. 20.4 Various designs of compliant grippers in their default and deflected configuration, as visible in the bottom right corner; PLA topology-optimized gripper has marginal deformation

Figure 20.4 presents four 3D printed grippers: analytically designed TPC gripper, the same analytical design of PLA gripper, topology-optimized TPC gripper, and topology-optimized PLA gripper. Both analytically designed grippers achieved the desired degree of freedom and have the expected kinematic behavior, as visible in Fig. 20.4. Nonetheless, it is important to mention that the TPC analytically designed gripper is prone to flexures buckling when there is force applied to output while input is fixed. Topology-optimized TPC gripper has satisfactory kinematic behavior, and as visible in the figure, it achieves the full closure of the jaws. In the PLA topology optimization, intermediate densities persist; it is not in the scope of this work to focus on this problem, but this feature had impact on the final design which did not have enough flexibility. In Fig. 20.4, it is visible that the maximum movement of topology optimized PLA gripper is marginal; according to FEM, analysis further deformation would result in plastic deformation. Nonetheless, the gripper will be considered in further modal analysis in the next section.

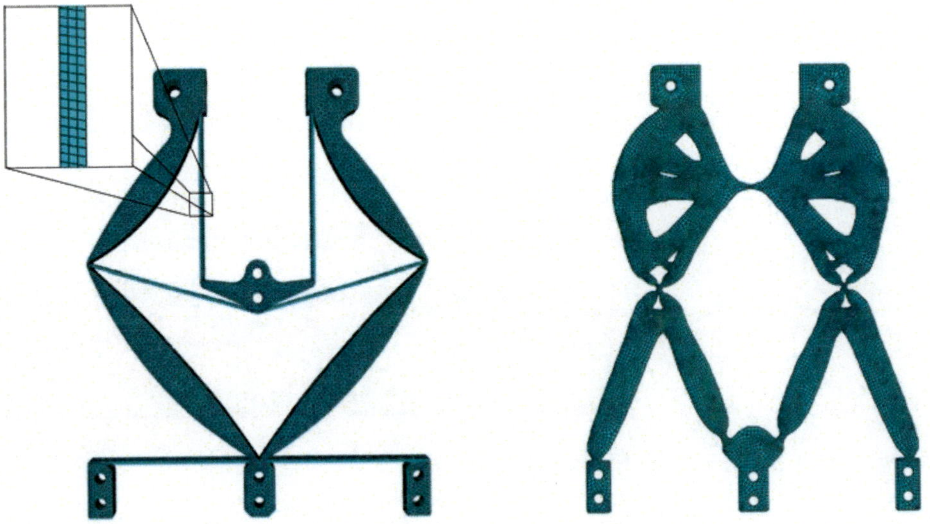

Fig. 20.5 Mesh of the designed grippers – analytical design (left) and topology optimization (right)

20.3 Dynamic Analysis

The dynamic analysis of grippers obtained using topology optimization and analytical design is here presented. The software ABAQUS by Dassault has been used to estimate the natural frequencies and the mode shapes of the designed grippers. The geometries of the compliant grippers have been develop using open-source software following the techniques presented in [15] in conjunction with Autodesk Inventor to generate STL files that could be imported in ABAQUS. Once imported, the geometries have been meshed using a mixture of hexahedral and tetrahedral elements: whereas the topology-optimized geometry could be meshed automatically, it was necessary to partition the geometry of the analytical design in several parts before a suitable mesh could be achieved. The reason for this is that the analytical design presents some regions with very high aspect ratio – very thin and elongated – which makes the meshing process more complicated. The elongated parts are generally referred to as "flexures" and are necessary to allow the mechanism to achieve the desired level of compliance. Figure 20.5 show the mesh used in the numerical analysis. It is worth mentioning that in the analytical design, an extra central constraining point is necessary to ensure the correct movement of the mechanism. Note that the holes in the tip of the grippers are used to facilitate the connection with external measurement devices and that they will be removed from the final design. For sake of accuracy, they have been reproduced in the model, but they are not expected to contribute to the dynamic response of the mechanisms, especially in the low frequency range. Both designs present four anchoring points in the lower part of the gripper to fix the mechanisms to the casing. This condition has been modelled using translational constraints ($U_x = U_y = U_z = 0$) applied to the internal surface of the holes. This is considered to be the most realistic model for the constraint introduced by the unthreaded bolt used in the assembly. The central holes are used to connect the grippers to the actuation system: by pushing (pulling) the central part vertically, the grippers are opened (closed).

Both the analytical grippers and the topology-optimized grippers have been manufacturing both in PLA and in TPC. It can be seen from Fig. 20.4 that the material mechanical properties do not affect the results of the analytical design approach while it has a profound effect on the topology optimization design. In this preliminary work, only one design for the analytical approach and one for the topology optimization will be considered to simplify the analysis and permit a fairer comparison. The geometry obtained from the topology optimization of the gripper in TPC has been considered for this work.

20.4 Results

A summary of the results for the grippers in PLA can be found in Fig. 20.6, where the first three modes of the grippers are shown. For each mode, the front and lateral view of the deformation is displayed, and the natural frequencies are reported.

A first point of note is that the natural frequencies of the analytical grippers are lower than those of the topology optimization. This is somehow expected since the topology optimization generally results in bulkier regions connected by very thin point-like flexures. This, in general, results in higher rigidity in the design and, therefore, in higher natural

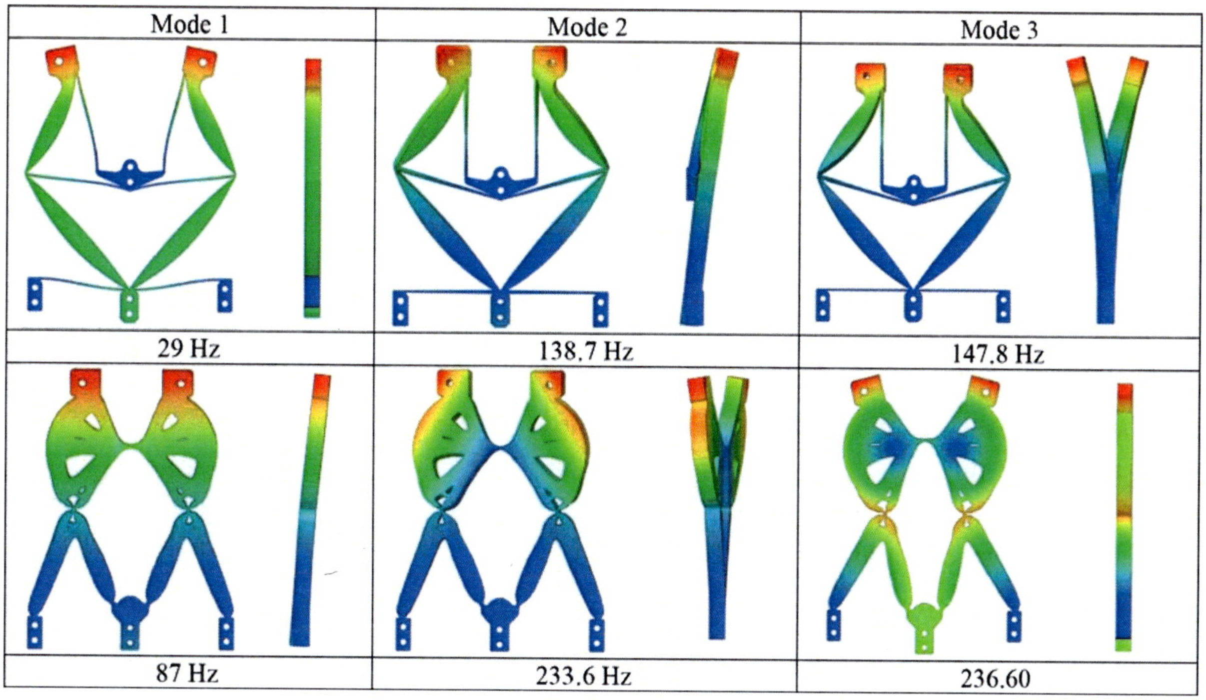

Fig. 20.6 Modal analysis of the grippers obtained with the analytical design (top) and the topology optimization (bottom)

frequencies. In reality, this highlights one of the limitations of topology optimization: the point-like very rigid flexures are also points where there is accumulation of stress, creating a potential weakness for this type of approach. The literature suggests that this problem can be overcome by using more complex optimization functions that consider the stress distribution in the mechanisms. To the best knowledge of the authors, no successful attempt at incorporating the stress distribution in the optimization function of compliant mechanisms has been recorded.

In Fig. 20.6, it can also be observed that in the analytical approach the first mode is fully planar and that the mode shape closely resembles the deformation that the mechanisms undergoes when it is used in its primary function. This is not true for the topology optimization design, where the mode shape that bears resemblance to the opening and closing of the gripper is the third mode. Bending and torsion of the body of the mechanisms occur at lower frequencies. This is further indication that the mechanism is quite rigid and offers considerable resistance to opening and closing the gripper. The resistance offered to the torsion and bending of this design, in comparison, is smaller.

20.5 Conclusions

In this work, the modal analysis of compliant mechanisms designed for extravehicular activities in lunar environment have been presented. Two design approaches have been considered: analytical and topology optimization design. A compliant gripper has been presented, which supports the astronauts in the collections of samples from the lunar surface. The presence of regolith, lunar dusts, causes significant problems to any mechanism that relies on hinges and slides between surfaces in contact. Compliant mechanisms are investigated for their capabilities to achieve the desired movements without the need of hinges. In this work, a preliminary study of the vibrations exhibited by compliant grippers obtained using the analytical and topology optimization is considered. The analysis demonstrated that the analytical design offers a greater flexibility in the desired direction and that the first mode shape resembles the deformation which is required by the mechanism in its normal operations. The topology optimization design instead offers less rigidity in out-of-plane bending and rotation, making the deformation required to open and close the mechanisms more difficult to obtain.

This is an interesting behavior that could carry some useful information for the design of compliant mechanism: the mode shape and the natural frequency at which they occur could be used to instruct the design process and result in designs that are more compliant in the direction of interest. This aspect will be investigated in future works.

Acknowledgments This work is cofunded by the European Space Agency (ESA) through Exploration Preparation, Research and Technology program under the grant number ESA AO/1-10811/21/NL/MG/idb. We thank ESA for the support.

References

1. Budzyń, D., Tuohy, E., Garrivier, N., Schild, T., Cowley, A., Cruise, R., et al.: Lunar Dust: Its Impact on Hardware and Mitigation Technologies. In: 46th Aerospace Mechanisms Symposium, p. 287 (2022)
2. Levine, J.S., Winterhalter, D., Kerschmann, R.L.: The Impact of Lunar Dust on Human Exploration (2020)
3. Gaier, J.R.: The Effects of Lunar Dust on EVA Systems During the Apollo Missions. Nasa/Tm-2005-213610/Rev1 (2007)
4. Pagel, B.E.J.: Physical properties of the Lunar surface. Lunar Sourcebook. **186**, 826 (1960). https://doi.org/10.1038/186826b0
5. Rickman, D., Street, K.W.: Some expected mechanical characteristics of lunar dust: a geological view. AIP Conf. Proc. **969**, 949–955 (2008). https://doi.org/10.1063/1.2845062
6. Colwell, J.E., Batiste, S., Horányi, M., Robertson, S., Sture, S.: Lunar surface: dust dynamics and regolith mechanics. Rev. Geophys. **45**, 1–26 (2007). https://doi.org/10.1029/2005RG000184
7. Gaier, J.R.: Regolith activation on the lunar surface and its ground test simulation. SAE Technical Papers (2009) https://doi.org/10.4271/2009-01-2337
8. Pirrotta, S., Lefebvre, D., Co-chair, M.W., Wong, H., Buffington, J., Gaier, J.J.: Dust Mitigation Gap Assessment Report (2016)
9. Howell, L.L., Magleby, S.P., Olsen, B.M.: Handbook of Compliant Mechanisms (2013) https://doi.org/10.1002/9781118516485
10. ADR, ESA, ESTL: Guidelines for Space Mechanism Ball Bearing Design, Assembly and Preloading Operations, p. 88 (2013)
11. Zhao, L., Yu, X., Li, P., Qiao, Y.: High-precision compliant mechanism for lens XY micro-adjustment. Rev. Sci. Instrum. **91**, 91 (2020). https://doi.org/10.1063/1.5141138
12. Pinskier, J., Shirinzadeh, B., Ghafarian, M., Das, T.K., Al-Jodah, A., Nowell, R.: Topology optimization of stiffness constrained flexure-hinges for precision and range maximization. Mech. Mach. Theory. **150**, 103874 (2020). https://doi.org/10.1016/j.mechmachtheory.2020.103874
13. Liu, L., Bi, S., Yang, Q., Wang, Y.: Design and experiment of generalized triple-cross-spring flexure pivots applied to the ultra-precision instruments. Rev. Sci. Instrum. **85**, 85 (2014). https://doi.org/10.1063/1.4897271
14. Koppen, S., Langelaar, M., van Keulen, F.: A simple and versatile topology optimization formulation for flexure synthesis. Mech. Mach. Theory. **172**, 104743 (2022). https://doi.org/10.1016/j.mechmachtheory.2022.104743
15. Budzin, D., Zare-Behtash, H., Cowley, A., Cammarano, A.: Implicit lunar dust mitigation technology: compliant mechanisms. Acta Atronautica (accepted, in press)